Learn H

Hypnotherapy

Double Edition: Volume Four & Five

Further Collected Works

&

Therapy in Action

By

Daniel Jones

GHR (Reg.), GQHP, DHypPsych(UK), D.NLP, HypPrac

BSYA (B.D, Cur.Hyp, H.Md, Zen.Md) MASC (Relax, NLP)

www.ericksonian-hypnotherapy.com

Second Edition 2011

Published By Lulu.com

ISBN 978-1-4475-7104-9

2 Second Edition 2

Other Books by the Author

'Advanced Communication Skills for Business Professionals'

'Conflict Resolution Handbook: Managing Violence and Aggression in the Workplace'

'The Psychology of Big Brother'

'Spiritual Teachings of an Unconscious Mind'

'Learn Hypnosis & Hypnotherapy: Volume One: Foundations'

'Learn Hypnosis & Hypnotherapy: Volume Two: Therapy, Techniques & Treatments'

'Learn Hypnosis & Hypnotherapy: Volume Three: Collected Works'

'Advanced Ericksonian Hypnotherapy Scripts: Expanded Edition: Over 100 Hypnosis & Therapy Scripts'

'Parenting Techniques That Work'

'Utopian Tao: The Secret for a New Generation'

'Take Control of Your Life: Self Help for Depression, Anxiety Disorders, Confidence, Success & More'

Acknowledgements

Thank you to those that showed curiosity and wonder by asking questions and looking to discover something new and interesting whether it was to help yourself personally or to help others professionally.

A special thank you to Graham Levell, Terry Stewart and Abbie Piper for constantly asking questions, not only provoking an answer and learning something new for themselves but allowing me the opportunity to enhance my own insight by going inside my mind to discover the answers.

Contents

Introduction to Part Four Further Collected Works

The collected works is a set of two books that make up parts 3 and 4 of my 'Learn Hypnosis & Hypnotherapy' Series. The aim of these books is to give more of an insight into my thinking and how I personally do therapy. Part one taught an underpinning to doing effective brief therapy, part two taught the actual 'doing' of therapy, parts three and four go into answering questions that have been posed to me over the years; questions relating to more fringe based areas of therapy from the use of energy therapies like Emotional Freedom Technique (EFT) to past life regression and future progression, to delving more into cellular healing and what research and information is available out there.

Much of the 'collected works' is edited from posts over the years in response to questions on forums and groups on the internet. Throughout the collected works you will find my answers to different situations that therapists find themselves in; like how to treat specific problems and difficult clients.

It is assumed throughout these collected works that you have a working knowledge and understanding of terms I use and techniques described. All of these are covered in the first two books. The aim here is to expand in a more free-form manner the information that you already know by giving examples and situations and ideas for thought.

Motivation and Paradoxical Interventions

My view is to always work where the motivation is, for example if someone is always being a 'class clown' what is the process and pattern. Often it is to avoid looking stupid, like having it noticed that you are not very good at something so you take control over the situation before you look stupid at not knowing something by looking stupid on purpose (although this on purpose may happen unconsciously)

I probably wouldn't use a paradoxical intervention I would want to know what the child feels they get out of their behaviour, what they would like to get out of their behaviour, whether they got what they wanted out of their behaviour, what alternatives they can think of in the future to get what they want etc...It's difficult to describe exactly what I would be asking as they may say they want to cause problems and they did so I would chunk up until I got to something acceptable (like fitting in or being happy) and work on that level. I wouldn't ask, suggest or say anything they could disagree with or resist so I probably would hardly mention the actual problem behaviour.

I would linguistically imply things and seed ideas and alter the future to the point that they think they have just decided to take a specific course of action...

Sometimes paradoxical interventions can be used even when there are family members etc...

If, for example, someone isn't sleeping then prescribing more of the same makes no real difference to others in the relationship as the person is up anyway. If anything the person will be getting up and going elsewhere so not being so restless in bed which can let the partner sleep.

Obviously there are cases where a paradoxical intervention may not be appropriate as it may have a negative impact on the family system.

With Children, when I worked in residential childcare with children with severe challenging behaviour if you tell them to stop swearing it wouldn't work, they would do it more, so often I would tell them to swear at me more and would encourage it and get them to try to think of more examples etc...they would get annoyed at doing as I want them to do and get irritated with having to swear at me more so they would stop...Obviously this would be done skilfully and respectfully so as not to increase aggression.

You need to think through the likely effect and consequences of the intervention. You don't want someone losing a job because of your intervention. With a child you are more likely to work with the parents as they are going to be the ones that want the behaviour to stop and so are likely to be more motivated. If in a therapy session you told a child to swear more for example the parents would probably be annoyed with this, what you want is the parents to be the ones encouraging the swearing (they will be contextualising it and making it very conscious and deliberate). Likewise if you have explained to the parents what you have asked of the child (like if you ask them to intentionally wet the bed - if they are a bed wetter) you want them to support you.

Sub modalities are a useful thing to know about. It isn't that you would create an unrealistic image of the future etc...changing sub modalities alters the memory (real or future etc).

If for example someone has to do something in the future but they are not motivated you can ask them whether they are associated or disassociated, then try on the opposite and ask what happens (increase or decrease motivation), then ask them to turn up the brightness and turn down the brightness and see what difference this makes. Do this for many sub modalities and remember which changes increased the motivation then use these so that the thought of doing that task is motivating.

Sub modality use is at the heart of many techniques, swish, Rewind, etc...Generally if something is brighter, more vivid, associated, in colour, HQ surround sound etc it has a stronger emotional content (not always, for example I have known people assaulted or raped etc in the dark that having the brightness reduced increases the fear), generally the opposite will reduce emotion, you only really know by testing or asking and paying attention to language. For example someone may say 'I just want to get some distance from my problem...' So you know they need to disassociate, or at least move further away.

Some examples of using paradoxical interventions successfully...

I worked with an alcoholic, his pattern was to go to buy three bottles of vodka, take them home to his flat and drink them there until he passed out, then repeat the process when he has sobered up enough to go to the shop to get more (normally the next day)

I had two sessions with him where he took no notice of anything, he had no intentions of changing yet his father had asked me to help him as he didn't know what to do...the man didn't ask for help himself.

It is easier to make someone that doesn't want to stop doing something agree to do more of it so this is what I suggested. I said he obviously want to keep drinking and wants to keep seeing me but I'm not going to waste my time seeing him if he isn't prepared to work with me. I said I don't want his father to think that this is a waste of time also so would he agree to do some drinking that will let him explore his issues with drinking and learn something and at the same time he can tell his father that I told him to do it as an experiment.

He agreed. I told him what I wanted him to do was when he feels he needs a drink, go to the pub, order 3 pints of beer, line them up in front of him and gulp back the first one then say 'f**king therapist making me drink this beer', then gulp back the second and third pints doing the same, then go home. If he still wants to drink more he is to go back to the pub to do the same.

He did this and stopped drinking on his own a few weeks later...(this type of intervention I have done on a few occasions with alcoholics that drink alone at home and that don't drink beer)

I worked with a smoker that was referred to me by the NHS smoking clinic. He came in and said 'I'm going to tell you what I told the lady that ran the clinic, I will have you smoking 50 a day before you can stop me smoking 50 a day'

He didn't come across as very motivated to quit, he knew all the reasons he had been told he should quit but said he didn't want to but he will see me and then he can tell everyone how he has tried everything.

I saw him for a first session, by the second session he stuck to his word of ignoring everything and wanting to stay smoking, so again I told him I want him to try an experiment just so I know a bit more about his habit, and that it will involve him continuing to smoke. He agreed he would do what I asked.

One of his patterns that happened every few days was that he drove to Tesco to get food shopping. He didn't go to the nearest supermarket because he had to have a fixed number of cigarettes. He would have one when he gets in the car, two on the journey and two in the car park (because he knows he can't smoke in Tesco so he has to have extra before going in there). I told him what I want him to do is to wait until he gets into the car park then smoke all five cigarettes and one more because he had to wait so long before having any. He enthusiastically agreed to do this thinking it sounded easy.

He cancelled his next appointment and I bumped into him in town a few months later. He said he was really angry with me. He did as I asked but he found that sitting in the car when he had shopping to do and *having* to smoke 6 cigarettes really annoyed him, all he wanted to do was get into the shop and get his food and go, but he was stuck outside the shop still smoking, he was resenting the cigarettes. He ended up deciding not to smoke in his car because of this. This then spread to forgetting to smoke after he was home out of his car and over a few weeks he was regularly forgetting to smoke and so quit.

Memories and Sensory Thinking

About what is going on in the mind...you have thoughts as a sensory experience as this is how the mind works, some memories will be dull others vivid etc...how the memories are coded in the brain influences the emotion of the memories, the memories will also be organised in your own learnt way of representing time...often in a line either left to right or behind you and out in front of you, you will often either be associated in this structure of time or disassociated, this also changes depending on the situation.

When for example the Rewind technique is done memories get recoded with altered sub modalities so that they are no longer traumatic. When Swish gets done memories get recoded with altered sub modalities.

Another use is for learning other people's strategies to take on for yourself, like if you wanted to think like Walt Disney and learn his creativity strategy you can take on his mental processes and the sub modality coding. Like when he was disassociated and associated, when he gets distance and closeness, when images are brought into focus and defocused, when they are bright and dimmer, when the volume level is changed and where the sound comes from etc..

Likewise if someone has a voice in their mind that says things like 'you're useless, you'll never manage that'...etc you can have them change the voice so that it makes them laugh when they hear it or so it evokes some other emotion that is more productive...

Another use with sub modalities is in pain control, turning the Kinaesthetic into other representational systems, like visual etc and then making the changes like increasing or decreasing light, changing the colour, shape etc...

Learning the Process and Structure Behind Techniques to Create Your Own

I am always promoting with people to learn structures and processes behind techniques so that you can understand what makes it work and then do your own in a way that suits the client not a way that you know how to do it...

Like learning about redirecting force in Aikido and about how this process is created in the various techniques so that you can respond in a situation with a technique you create in that situation utilising the processes rather than trying to think of a technique that might work...

Relaxation; Trance and Trance Signs

Soldiers go into a trance marching and are completely not relaxed.

I have worked with people with things like phobias in the situation they are having the phobia where there is no way of relaxing them or doing formal hypnosis of any sort, because the person could be at the top of a tower about to abseil or about to hang onto a death slide; so they need to be helped in minutes.

Often people are deeper than they first seem, and when you watch them you notice the trance indicators even though they seem alert and awake...

An induction I like using is to guide people deeper into a trance by using a painting in an art gallery in their mind; or a room with a TV or Cinema screen they can step into followed by other paintings or screens in those paintings or screen that can take them deeper and more fully absorbed.

Another induction that is less structured word wise and offers more freedom for creativity for the therapist with the sensory language input is to use a journey or adventure with multiple routes to take. You could use a holodeck or the adventures of Alice in wonderland with doors and rabbit holes, twists and turns etc all leading to new discoveries. I have used people settling in a situation like by a flickering fire and drifting into a meditation where they can find themselves in an experience where they see themselves meditating under a tree perhaps, then they can lower into that them and discover where that them is meditating about etc...I think it is useful noticing processes not just taking a technique and using it as it has been learnt.

I feel the more out of the process the therapist keeps the better, so unless a client mentioned holodeck or star trek I wouldn't do that, I wouldn't tell them where they are meditating etc...

In my experience you don't need the person to appear to be in a deep trance to have effective therapeutic work take place. I have had people either too hot or too cold or uncomfortable and they have moved or opened their eyes and got sorted out without any real negative effect on the trance or the work done.

People go in and out of trance with each sense system. At one point they may hear a noise so their hearing comes out of trance whilst the rest of the person stays in trance, then when they are comfortable with the noise their hearing goes back into a trance again, then if they need to open their eyes their eyes come out of trance yet their hearing and feelings are still in trance, then they go back again when ready. Likewise if they are uncomfortable how they are sitting they may come out of trance to shift position then back into trance again.

I have experienced this in many situations, even when people have had to get up completely because they hadn't turned off their phone and it has rung so they have come round, answered it and then afterwards sat back down and as soon as I have continued talking in the same hypnotic way they re-enter trance almost instantly.

When I started out I used to worry that people had been asleep, not hypnotised. I had regularly been reassured that it is incredibly rare and that in the rare occasion that it happens it usually mean the person needs the sleep more than the therapy on offer. Nearly every person I hypnotised told me they were asleep because they don't remember anything. I learnt that they weren't because they always opened their eyes on cue.

It became apparent that these people's unconscious was listening and that the conscious had gone off somewhere else...The unconscious always responded in a way that let me know it was listening. I used to test if I was unsure by getting the client blushing on one side of their face (as a visual cue not easy to fake that if they are sleeping they are unlikely to spontaneously do)

I used to get hung up on having to get people in a deep trance. Now I am so unconcerned as I know the unconscious is always listening, most of what I do probably doesn't resemble hypnosis. This is something Steve Gilligan said when talking about doing Deep Trance Identification as Milton Erickson, he said that one thing he learnt was that everybody's unconscious was listening and that everybody was already in a trance.

Polarity Responders

My experience with polarity responders is that they are everywhere. Children and Teenagers especially seem to be polarity responders, you tell them to tidy up and they make more mess, you tell them to be quiet and they turn up the music etc.

I think it is situational in many cases; they won't respond well to authority in one situation but may do in another situation...like bad with the police, but good with peers...

Everyone probably is in some situations where if someone tells you to do something you will do the opposite...Some people are more this way than others...When this happens with the therapist it could be that the therapist represents something (like an authority, or a person from the past or a father etc...) or they are annoyed with the situation, they may also be desperate to comply yet respond this way...

Generally I've not had problems with polarity responders, it is more difficult with people that completely ignore all options as you have to think in detail to make sure you have covered all options leaving them with something that wasn't an option that is what you want them to do but that if they decide to try to catch you out and go with one of your options these will also lead to the outcome... I think this 'non cooperative' person isn't a polarity responder as they don't respond opposite they do their own thing instead.

A true polarity responder will respond with the opposite response. To your question 'which hand they feel is heaviest?' A polarity responder will say they feel light as saying either hand will agree with the question. By saying a light hand they are agreeing one is heavy, if they say both are light then you would have to utilise this.

By saying what you don't want or saying negatives you can get round the polarity responder.

The Classic Staircase Induction

One issue with the staircase induction is matching people's internal reality. When you are timing steps with each out-breath it relies on the client also going at that speed. Even when you say that it can be one step with each out-breath people sometimes find this too slow as they are walking down the steps and they are already at the bottom as you are counting.

Other people follow it well and go at the speed of their breathing.

If people seem to be rushing to the bottom of the steps some things that I have found useful in slowing them down is things like having them wonder what is around them, pictures on the walls, views out of the windows as they walk down the stairs. Or just wondering whether they will get distracted as they reach the 6th step 7th step or 9th step and wonder what will distract them...etc

Alternatively I separate the counting and the steps so they can walk down a flight of steps and it will take them until the count of 20 to reach the bottom.

I'm one of those people that wants to go at the speed a hypnotherapist is saying but am often at the bottom waiting for the counting to catch up, or I'm running up and down the steps as I wait...

Another way is to say something like 'I wonder whether it will be walking down the step or the counting down that takes you deeper into a trance or if it will be the breathing out with each number that helps you to become more fully absorbed'

Why Do People Have a Built-in Process For Getting Phobias and How Can You Help Them?

The feelings are required for survival purposes so the memories can be repressed but the feelings and the response to future situations has to come through.

If you were attacked by a sabre tooth cat (in prehistoric times) you may get scared enough to cause a phobia. In the future under similar conditions something may happen that evokes the fear response to save your life because it could be a threat. You don't have to know what the initial incident was only that in future situations you will instinctively know to be afraid and run or fight.

This process has saved people for thousands of years.

Treating Phobias

Use Scaling: The difference for the client is more noticeable between scaling before treatment and after treatment. Phobic clients go into a trance easier. There is more to work with. Trauma memories are stored as survival memories so the more phobic someone is the more pure the memory. You want to change the memory to being stored like a normal memory. If they can't access it so easily it gets harder to work with.

Use the rewind technique or a variation of this: The reason for rewinding the memory is to scramble the pattern having it run from after everything has calmed down to before anything happened, these are also two safe points

I usually do it about 3 times, it could be less sometimes, sometimes it maybe a couple of times more, not often though...

You can ask them to re scale after the work has been done to notice the new reaction...

My view is that if after treating one memory if there is still a bit of a reaction there then I go for a second memory and occasionally a third...

It is called Rewind because that is a part of the technique that is done, it is called Fast phobia cure because it can be used to cure phobias fast. Although obviously it isn't exclusively able to treat phobias, it can treat PTSD, OCD, etc...

The Cinema metaphor is just one metaphor for the process...

As an example I was sat on the plane back from Dubai and the person next to me was nervous about flying. I picked up the in-flight duty free magazine and flicked through it finding an image of a child playing with a children's airport play mat and big fat comicy looking aeroplanes. I got all excited about playing with the plane, flying it all around the air and the airport etc...

The person found they got excited about the flight and enjoyed the flight...

I used principles of the Rewind Technique but used the image of the child playing with the plane and airport and changing their representation of flying. I used language to link what I was using with them. It took about two minutes and they could think of flying with enjoyment rather than fear...

With these techniques it is about learning the process. Once you know that you can improvise your own treatments.

Once the memory goes into the higher parts of the brain future situations will be processed more appropriately, rather than through a strong emotional filter. It doesn't stop the person being able to have the fight or flight response to certain related stimulus just makes it so that it is appropriate. For example; having the fight or flight response in a future car accident is appropriate, having it every time you see a car because it reminds you of a previous accident. Or having a fight or flight response in a situation where someone is trying to rape the person is appropriate, having it when seeing any man or any specific colour just because it reminds them of a past incident isn't appropriate (for example I worked with someone that had been attacked and raped by someone wearing black in an alleyway. The person developed a fear for black clothes, alleyways and men behaving in certain ways. This incident was de-traumatised so that none of these things triggered fear, if someone attacked her again she would still respond appropriately for her own survival).

By survival memories I mean fight or flight response memories. They are encoded with the more primitive parts of the brain that deal with emotions predominantly (Amygdala). By taking the emotion from the memory(ies) yet keeping the content it gets stored with a lower emotional level of arousal and so gets stored in higher parts of the brain.

This isn't just fight or flight emotions, it is all strong emotions. Fetishes are also stored in the same way as are other very strong emotions. That is why if you do the Rewind Technique on someone's fetish it reduces the strength of the fetish, and why you shouldn't use the technique on a pleasant strong emotion like love.

I worked with an agoraphobic a while ago, she said she had been agoraphobic for many years and that it hadn't been a problem because she just stayed indoors, didn't go out, had the internet to do shopping, had people round etc... It had been a problem until she decided not to try going out anymore.

Her problem started with walking down a pavement when a cyclist knocked her over, she got knocked out, hospitalised and nervous on pavements because the same thing could happen again. A few weeks out of hospital and she was in her car when another car hit it. Again she got traumatised and felt if it can happen on one road at one time it can happen on any road anytime. A few more minor bits happened just so that she could confirm and maintain her belief before deciding it was safest indoors. Her opinion on 'safest indoors' was that there was no problem when she was indoors so it was like she had no problem (just like someone scared of spiders has no problem in a confirmed spider free zone so they could become complacent and almost believe the problem doesn't exist)

I spent some of the first session de-traumatising her past experiences and doing other bits and pieces, then we went for a walk down the beach and had an ice-cream and just chatted. She had not been able to go outside in her garden or out the front. The only thing that made her 'remember' she had a problem, or realise the problem was that her son was in hospital and she wanted to visit him and wanted to be able to bring him back to hers to look after him and found she couldn't leave the house when she tried.

I had a few subsequent sessions to get her driving again comfortably...

I have used the Rewind Technique in hundreds of situations effectively. What happens is that the person is double dissociated at first (so there is virtually no connection with the feelings), they watch them watching the movie of the old incident, then they experience it rapidly in rewind, this scrambles the pattern as they experience it in a new and different way (everything running backwards, often sounding stupid because all audio and voices are fast and backwards), then they dissociate again and watch it in fast forward (again they are watching it in a new way which scrambles the pattern again), this process happens a few more times before they watch it at normal speed again. This is a case of using time distortion and dissociation to turn the old traumatic memory into a normal memory. Traumatic memories are accessed differently (because they are for survival) to normal memories.

How Do You Get To The Root Of What A Client Really Wants From Therapy?

Often the easiest way is to ask 'if you were to see yourself on a TV screen once the problem is gone how will you know? What will you see, hear, feel? What will others notice...etc' all of this gets information about what will be achieved rather than what will be gone.

It is often difficult for people to know what they want rather than what they don't want (like with smoking etc) by asking them for a description of what it will be like instead of just asking them what they want people often find it easier to answer and it gives you information to use in the session.

I rarely write during a session. It often just seems like a chat; my aim is to keep the therapy relaxed and conversational. The questions I ask would be What, Where, When, Who, How questions. You would want to get sensory specific information, one of the easiest ways to get this is to ask what would be seen if you were watching it on a TV screen. People say things like 'I walked out on stage and the panic hit' this doesn't give information on what happened physically, linguistically or internally (internal dialogue, body feelings). It would be useful to know at what point did the panic hit. How did they know they were panicking not just excited? What lead up to the panic (obviously you would not want to use the word panic as often as I am here!)(Did they spend weeks worrying about panicking practicing for the real event?) Who else is involved (does the group have to be over a certain size, made up of certain people etc) the other option is to ask 'If I were to give you a day off and I'll do the panic for you, how would you teach me to be able to do a good panic, up to your standards? (Then they can tell you to worry for weeks, through to the actual event and then how they calm down afterwards)

Creating Dissociation, Metaphors and Age Regression

My view is that a metaphor is a form of dissociation. The metaphor maybe used to give the client a pattern for the necessary association, for example if I told someone about streams running to a river then the river running to an ocean it would lay down a pattern for association or integration, the *result* of the metaphor would be association. Likewise if I told a story about a cat that lost its kitten and blamed herself so couldn't cry or express emotion and spoke about the cat's experience of getting in touch with the feelings, that would be laying down a pattern and would be dissociated from the person and their actual situation, hopefully getting in touch with those feelings would happen as a *result* of the metaphor.

Likewise with smokers I would often discuss paths in forests and cutting through a new path that once cut through sufficiently is easier, quicker and more pleasurable to follow. Or about someone moving from a smog filled city, with all the congestion on the roads etc...to moving to the country with open space, fresh air etc...

Even metaphorical tasks have the person dissociated as they are experiencing a pattern (like going into a field to find two identical blades of grass) that will be of use to solving the problem, they aren't actually experiencing the problem itself, or the solution itself.

Using the Crystal Ball technique

I had a client I used this technique with; again rather than the standard version I created my own version based on the principles.

I started by talking about how the sun is 8 minutes away and how what you are actually seeing is what the sun looked like in the past and how you can't see what is happening now. I then mentioned other planets and stars and how each of these is also being viewed at different times in the past, that when you look at the night sky you can see many stars and planets and the moon simultaneously yet the moon is what it was like seconds ago, the planets are what they were like minutes ago, the stars are what they were like years ago...

I then moved on to talking about how as a child you get so engrossed and focused on cutting exactly around the shape of a person on folded up paper, and that the more you cut the more careful you are to make sure you cut it really well, and you focus completely on that paper and on that cutting...you can't wait to see what it will look like...you're excited to find out...when you have cut out the shape of a person completely you then slowly and carefully open up the paper to have it look like lots of people all holding hands...

I then mentioned how each person can carefully be coloured in so that each one looks slightly different from the next, with a youngest one on the left and the oldest one on the right...and how each one can be coloured in to represent a relevant part of how the issue we discussed (it was smoking) was able to form, formed and had been able to be maintained up until now...

Then I spoke about how many great works of art have many layers where they have been changed, updated and corrected until the artist feels that the picture is right, and that they will take all the time they need to make it just right...

I told them they can continue to slowly and carefully colour in each person until each person feels just right...and how they can get a sense of how colouring in one person can influence how the others need to be coloured in and altered...

I had them do this until they were proud of their work and could step back and admire the end result (head nod to let me know this was done), then I had them carefully fold back together the paper noticing how it can become more 3D as each part is stuck back in place with the newly painted images integrating in their own unique way...

I moved on to some more stories before moving back to talking about space, stars and planets then the sun, and then allowing them to open their eyes when they have fully reintegrated in anyway necessary and made all the changes needed to allow progress to be automatic and to take effect at an appropriate rate and speed...

Another way for inducing age regression can be to use Double Dissociation Double Binds:

'You can drift into a pleasant memory and wonder what the future will hold, or discover yourself already in the memory curious about the future'

'You can experience a pleasant memory with no awareness of the future, or be absorbed in a memory looking forward to the future'

'That memory can take you back to a previous pleasant experience before the future happened, or that memory can take you deeper into the past curious to discover the future'

When I was out in Dubai recently I found that, being from the UK, I wasn't used to all of the heat. I would be beside the pool, lying in the sun and after a while I would get used to it even though at times it was uncomfortably hot. When it got uncomfortably hot I would go to the pool and go to walk in. The temperature difference between being out of the pool and being in the pool made the pool seem much colder than it really was. It made it difficult to enter as it felt too cold and uncomfortable. I had to decide whether I want to be hot and uncomfortable or cold and uncomfortable. I knew that staying out of the pool would get hotter and hotter as the day went on, and more and more uncomfortable, yet also knew that once I was in the pool I would be fine, it was just taking the steps to get in the pool that was the challenge. In the end I decided 'sod it' and just jump in, and quickly got used to the water and feeling comfortable...

I told this story (not exactly the same, I tailored it to the client) to a client the other day. They were depressed, they seemed proud of how many Psychiatrists, Psychotherapists and Counsellors they had been to and that they had spent time in the priory and yet they were still depressed. They explained how they will 'always' be depressed so they 'have to get used to it' and that they were told they should see me just to 'talk it over'. They said they are uncomfortable with change and have tried CBT with no luck because they know what they should be doing and saying and they know that what they stop doing when they are depressed are the things that will stop them being depressed but they can't put all that into practice once the depression starts. They knew I had just got back from Dubai and so asked me how it was, what it was like (which is why that story came to mind)

The following week the client was much happier and cheerful (still has bits to work on) and she was using terms I used in my story to describe how she has been (like 'I decided 'sod it I'm just going to go for it' and feel uncomfortable mixing with people when I'm down because I know that will make me feel better')

The metaphor I used above is one I chose to use because it is a true event from my life that I can tell in conversation without it seeming like a metaphor, I'm just talking about my holiday experience. I find that the most important thing is to have a thread running through the stories you tell. So if you wanted trance you may talk about interests and as you talk about your own and the trance aspects of them (without mentioning trance if you want to be indirect), then you may talk about science (if you or they show an interest in that) and fascination with Newton under the apple tree and Einstein day dreaming travelling on a beam of light, you may then end up on the subject of holidays and so talk of trance aspects relating to holidays, etc...All these stories will make sense in context (EG;

discussing interests, holidays etc) with what is being discussed, also they are being discussed in a wider context of the overall discussion about the perceived problem so they will unconsciously make sense in relation to the problem. As each story has a same pattern in it (that of people entering trances spontaneously and effortlessly and positively etc...) the unconscious mind can spot this same repeated pattern as it is in each story. The same with hypnotic phenomena or patterns for resolving problems etc.

You can also be vaguer with patterns in stories especially when someone is in a trance, like stories of nature, seasons, animals, fairy tales, etc...

The unconscious is very good at working with patterns, so if you created a metaphor that laid down a pattern it can use that pattern in a different context (the problem context).

I worked with a French girl once that barely spoke any English and would have struggled to understand the words I was using if I used complex language patterns and may not know half the words, she wanted to quit smoking, she could speak some English so we could establish like and don't like. This was enough to start working with, the rest was images, holidays, demonstrating deep breathing in and filling lungs, not liked places, not liked images, shallow breathing and coughing and suggesting she should visit Arundel (a local countryside town), go to the top of a hill and breathe in some of that fresh air and wonder what it can mean in the context of being healthy. She stopped a few weeks later after doing all this and we had just enough language to get by...

Hypnosis, Trance Induction & Utilisation

One quick way to induce a trance is to have a person recall their problem (it is often likely to be trance inducing), like getting a smoker to recall smoking (or getting a craving), or a person in pain to focus on the pain (only this time in a non-attached way be focusing on its colour, shape, size, etc), or a person that has OCD to discuss their OCD process, or someone with a spider phobia to recall the phobia, etc...

The higher the level of emotion the deeper the trance the person will naturally go into when they recall it.

You're always working with the trances you get, some people are just more responsive than others and so better hypnotic subjects.

Everybody is different, some people you can just look at them and say sleep and they will (if they know you do hypnosis and expect it to happen). Others would not respond in this way.

A good hypnotic subject is likely to be able to perform hypnotic phenomena and respond to therapy easily.

As Erickson has mentioned, in some cases he had to train people for some time to help them to be good hypnotic subjects. It is useful to know when someone is at that stage, so that you can move on to hypnotic therapy using different phenomena and so that you know they will be more responsive to what you say, whether this is when you first meet them or after you have trained them for some time. Generally though people don't need to be brilliant trance subjects to do good therapy, the therapist just needs to be able to utilise whatever the client brings to the therapy.

I naturally take fairly unnoticeable long slow breaths and people think I'm mucking around and holding my breath, this is often (not always) more pronounced when I enter trance. If I am hypnotising someone it is my responsibility to make the effort to match the clients breathing.

The trick to breathing quicker (but slow for the person your matching) is to just drag in and push out the air at a faster rate rather than do half of your normal breath then breath out (always leaving half your lungs full of stationary air because you never empty your lungs properly) as not emptying your lungs properly and filling them properly is bad for you. It is a bit like scuba diving and having to learn to control your breathing, then after a short while you can do it automatically.

If someone is breathing too fast or in a way that would be awkward then don't copy it exactly, you could do 3 of their breaths to one of yours (or any other comfortable option). And you could make emphasis to the out-breath and may be do your in-breath to 3 of theirs, then your out-breath to 4 of theirs.

There can be so many contexts when you want to notice as people enter mini trances so that they will be taking on what you are saying (assuming the trance includes you) or they could be in a trance to integrate what you have just taught (like doodling or staring into space) so you would want to give them a brief bit of time to finish. Or if you want to demonstrate and have as few problems as possible then someone very responsive is likely to carry out what you say best (which can also act as a convincer to the less engaged)

If you ask someone about the stages of their problem they have to enter trance to tell you. If you ask them about a leisure activity they enjoy they will enter trance. If you ask them what colour their front door is they will enter trance. Ask them how they will know when they are better and they will have to enter trance...

It would be difficult not to have them enter trance. Even if you sat doing nothing they will go inside to ask themselves what is going on, so they will have entered trance.

These are all small and can be built on and used for a bigger future trance, or any of these can be deepened as they appear.

When you ask someone 'have you ever been in a hypnotic trance before?' what you are doing is a double bind. This is because you have added the word before. If you ask have you ever been in a trance? They can say yes or no, if you ask 'before' it means before what? Before the one you are in? Before the one you are about to go into? So whether they answer yes or no they are accepting they will go into or are in a trance.

If they answer yes and it is a good experience then gathering information will quickly drop them into a trance again yet it will appear like you were just enquiring about that previous trance. If you want to still follow this line of questioning to induce trance when they have said no you can just explain what it will be like (using your hypnotic language skills)

Either way they are likely to enter a hypnotic state rapidly and be well on their way before they know what is happening.

I just wanted to share my experiences of stopping using scripts.

When I first trained everything was direct and all about using scripts. I even contacted every therapist in my area to learn from them, get their opinions and views on their success etc and all the feedback was to buy lots of scripts and when a client tells you what their problem is, use a script for that, find out which induction script they want and use that and use a script for ending the therapy. I had a collection of over 500 scripts! Imagine sitting with a client and trying to remember which script I should use!! I also felt it was wrong to just read in a monotonous voice from a sheet of paper and get paid for it and claim I know what I was doing. They could buy a book of scripts, choose the ones that suit them best, talk to a tape machine and do it themselves for much cheaper.

When I found out about Ericksonian Hypnosis I realised what Stephen was doing and Richard Bandler and that it wasn't that they had memorised inductions and therapy scripts and were reciting them, but that they were tailoring the therapy to the client.

I attended a two day course on Ericksonian Hypnosis and on the course we had to sit opposite someone and (like catchphrase) 'say what you see'. This was fine and I was comfortable with this in the safety of a course where at least I know I could do hypnosis, there were beginners that couldn't. I had also by this point started 'ad-libbing' self-help tracks because I couldn't find tracks or scripts for what I wanted to explore. I had also listened by this point to many of Stephen Brooks' Audio courses and seen numerous videos and so had a greater grasp of language patterns, tonality, etc...I still used scripts with client because I thought I would not know what to say.

After the course I met up with a friend that was willing to be a guinea pig, I said confidently that I can now do hypnosis without a script. I decided I would do a leisure induction with him and utilise his interests and times his mind has naturally wandered, and utilise on-going behaviours that I can observe.

I asked him 'in an ideal world where you could do anything, what would you do that would make your mind wander, that would make you lose track of time and really enjoy yourself?'

His response was 'I would go back to Thunder Mountain (apparently some water-ride in a water park in America?)'

I thought well I said I would use anything...so I did, and he said it was the deepest trance he had ever been in and we got numerous hypnotic phenomenon and great success.

I was nervous when he didn't say a nice warm beach or something like all the course participants had said, but I am glad, I have never looked back and now can't imagine using a script.

The thing I learnt is you can't be wrong because you are given your script moment by moment by paying attention. And if you expect them to go into a trance and so let your voice and breathing guide them it doesn't matter if you don't yet know all of the language patterns. You learn best by being uncertain at first rather than knowing it all then deciding to try it out.

There have been a few occasions where I have worked with people that need to know the side effects of everything. You talk to them and they tell you all about all the different tablets they take and how they always get most of the side effects. With these people on many occasions I have got them to be agreeing that when they receive treatments they have the side effects. I then give them side effects for the treatment they receive from me. These side effects are obviously positive though.

I do this when working with some people with Obsessive Compulsive Disorder also. I will give them a daily treatment plan that sounds specific but isn't, like between 1830 & 1945 you will have fun with your children, the plan gets followed obsessively, I have symptoms created of what happens if the plan isn't followed (positive of course) that gets the person trapped in a double bind. Doing the re-framing and getting agreement initially is the trick, once they are willing to follow the plan they also tie themselves into following the consequences of not following the plan...

Describing your own experience to induce a trance

You know one of my interests is going on walks through the nearby woods. I'll spend hours just wandering along in my own little world...feeling the breeze on my skin...I...begin to notice the sound of each footstep...time seems to just... slow right down...and I seem to be able to ...notice the smoothness of the movement of my breathing, of each regular step, of individual sounds from the birds, the rustling of the leaves...noticing the shimmering rays of light...the warmth of the sun on my face...and as I continue walking I...notice how the breathing begins to relax and deepen all by itself...often I find my...muscles relaxing...around my shoulders, arms, neck and face...and before long it already seems like time to go home...

I find when I talk hypnotically about an interest I have the client often finds it a familiar experience and so gets guided indirectly by listening to my description. I did this for one person (a hypnotherapist) where I challenged myself to see if I could hypnotise a hypnotist without them noticing. Part of what I did was said 'you know I've always wanted to drive down America, see how things change on a journey through the States' I went into detail about this imaginary journey in conversation and he was in a trance in no time at all.

Regarding therapist entering trance as well as client, I agree it is best when it happens. The difference is in the trance. The client enters a trance focused internally and the therapist (at least in my case) goes into a trance focused intensely on the client, paying full attention to the client. So the therapists trance is an externally focused trance, the clients internally focused.

Sometimes the therapist may not know all they need to know or they may not have time but want to get as much done as possible or they may have been presented with a number of issues and only worked on the one they could make change the fastest.

Nominalisations can be used to aid the client's unconscious to begin to spread change to other areas. I have recently posted a video on a site of mine where I work with a woman that over the phone said she wanted to quit smoking then came to me and said she wanted to lose weight and stop drinking cola and quit smoking. I asked which one of these was most important to her. Quitting drinking cola was what she expected to find hardest and was most important to her. I helped her with this issue whilst dropping in nominalisations and non-specific ideas for change in other areas to also occur. So far (three months later) she has lost about a stone and a half, cut down on smoking and had no problem stopping drinking cola with no side effects. She wants to now work exclusively on smoking in a follow up session. My aim was to promote a way for her unconscious mind to have permission and an understanding to spread change. Asking things like 'You can be curious to discover what other changes occur' A sentence with no specific meaning other than the one the listener places on it and it doesn't give any direction or content as to what is expected other than change. Given in a context where all change that is happening is positive the expected change is also likely to be positive.

In one session I couldn't cover all three issues but could indirectly begin to get movement on the issues I appear not to be working on.

Arm Levitation and Catalepsy

Just lifting an arm in an ambiguous way would induce catalepsy without asking for it (the movement would imply it).

Saying 'In a moment I'm going to lift your arm and I'm not going to tell you to put it down' Implies you are going to want the arm to be cataleptic but doesn't say this.

Telling a story about being in a cinema and your hand stopping in the air as something interesting happens on the screen implies catalepsy.

Saying 'and when I lift your arm you don't have to move it up, or down or left or right or in any other direction, you can just enjoy the relaxation' (implies it will stay still)

Talking about animals that lie in wait for hours on end without moving implies catalepsy.

The stories or metaphors above would be useful for seeding in advance, giving time for it to sink in then when you lift the arm the unconscious recognises the pattern and activates what was seeded earlier.

Another way could be to lift the arm so gently they client doesn't know if you are holding it or not so it gets confused and stays where it is.

Time levitation instructions/suggestions/commands with the clients in breaths...

It is a good relatively easy form or ratification keeping the arm in catalepsy when the person awakes from trance. I think it all depends on the person and the situation and what you want to achieve with it...

That is catalepsy, catalepsy is happening all the time somewhere in your body (like the neck staying in position to keep your head still). Catalepsy is not rigid like an iron bar (although this is often used as a metaphor) it is more like a waxy immobility that is comfortable, it is difficult to describe but your description is correct.

I have done full body catalepsy in un-hypnotised people by having them stand and then tapped on their shoulders in different directions causing confusion (like the tapping on the arm) and on the upper body and catalepsy sets in.

At the same time it makes the areas reduce in ability to feel sensation, they also stick where they are placed (if you lift a cataleptic leg it will stay where you let go of it, for example). It is good for initiating pain control or for operations.

In catalepsy there is no muscular forcefulness/tension, for example if the eyes are cataleptic it isn't like they are being held shut like when you tightly shut them, it is more like they just don't work.

I worked with one person that needed to believe he had been in a trance and I was videoing the session and I asked what would make him believe, he said if he could see on the video that he was in a trance, so I had him do catalepsy for the whole hour...He was convinced because he knew this was impossible normally, his arm would have wavered or lowered.

Other times I see that someone's arm is getting tired etc so I will suggest faster lowering, sometimes slower lowering, sometimes if they believe 'I have the power' and I need to be a bit more direct I wait until the arm is halfway down then I push it down to their leg as I link it to something internal almost like a shock/surprise induction being done with the person already in trance.

Other times I use it as a metaphor for something so it could be that I can lower the arm, then they can lower the arm (like lowering a resistance etc)(with the arm placed between me and them)

Hallucinations

Suggestions can still be given indirectly, or priming/seeding can be done indirectly etc...

If someone doesn't see what you want them to see then you can reply with something like 'that's right, you really don't see it, and I wonder what else is there that you really can't see...'

The other way round it is to presuppose what you want them to hallucinate without saying it, like asking 'what breed do you think my dog is?' Whilst slowly gazing down towards the floor where you want them to hallucinate the dog. Or what do you think of my new picture? Whilst looking at a blank wall. If they say they don't know they can't see it cause confusion by implying not seeing it means being deeper in trance, and praise them for their ability to go so deep, and then deepen their trance etc...

If it was auditory hallucination you can mention how you can hear music in the background and ask if it is a piece they are familiar with, and then ask them to really focus on that music.

With hallucinations in most positive hallucinations is a negative element and vice versa, for example, if you hallucinate a chair in front of you; you have to hallucinate out parts of the background, if you don't see a chair that is there you have to hallucinate in parts of the background.

I once decided to do an experiment involving creating artificial auras. I hypnotised myself to see different colours around people for different modes. My logic was that there is so much information to take in (non-verbal signals, verbal cues, words, etc) that I thought my unconscious is probably noticing all of this stuff, can it just process it for me and give me a cue that I can notice that sums up the information. What I thought was the best thing for this would be to see auras that I can observe changing and can work with (for example if I wanted a specific depth of trance, once the client is there the aura would be a dark blue, I this gets lighter they are coming up so I need to acknowledge then deepen, if it gets tinges of red there is some anxiety so I need to acknowledge this and deal with it etc)

Over time the auras faded and I just started saying what came to mind whether it made sense or not.

I used a similar thing when I first started out doing hypnosis when I was about 14. I was envious of people with synaesthesia I thought 'if only I could see sounds, how useful would that be for playing manhunt in the woods at night'. So this is what I did. I made it so that sounds would make light and so if someone stepped on a twig for example I would see a flash and know where they were. It is easier to judge where something comes from when you see it rather than when you just hear it. This is something that frustratingly I've not been so capable of as I've grown up.

Surprise and Confusion

Surprise or confusion can be used indirectly, just telling a joke can surprise. A handshake being slightly different is barely noticeable or paid attention to but it causes confusion and a trance, moving your head as you talk to the client (or looking into a different eye for conscious/unconscious etc) causes some confusion as two messages are being conveyed with different meaning, one of them to the unconscious to go into a trance, overloading the client with information causes confusion, like asking them to do something then before they have time to get it done ask for more and more until they need to take on some of the tasks unconsciously. making purposeful mistakes can cause confusion like saying I'm going to reach over and lift up your left hand' and lifting the right one, or saying different to what you are doing 'I can lift your hand up, down left right etc (doing as you are saying) then after a few rounds of being congruent move the arm different to what you are saying (up, move hand down, down move hand up, left, move hand right etc)

When I tell confusing stories changing terms with one meaning into characters people seem to think it is a challenge to do. The trick is to turn each term into a character then just tell a story, it isn't confusing to me saying it (unless I rattle it off too fast) because it is just a story and characters.

Left/right confusion etc is all something you go into a trance to do. I normally do small doses not long reams although I don't have a great problem with this. It has to be right for the client and context.

Client says left hand feels heavier than right, I might say so your left hand resting right there is heavier than the right land left right here is that right? Etc...

Compound Suggestions

Compound suggestions can overlap. Generally it is a truism followed by a suggestion, this can be from observable to non-observable, out of trance to in trance, etc...

For example:

You can sit there, and read this writing

You can read this writing, and let thoughts come to mind

Those thoughts can come to mind, and some can be of pleasurable experiences

You can be aware of those pleasurable experiences, and become more absorbed and relaxed

One thing I did when initially learning this and all of the other language patterns and structures etc was to listen to conversations (in real life and on TV etc) and look out for specific patterns.

In work lots of times people would say things like 'Your shift doesn't finish for another hour, does it? Can you go get the paperwork up to date' Implying because the shift doesn't finish the person can do the paperwork although there is no real link between the two.

In ordinary conversation people don't often work from observable to non-observable, or from not in trance to in trance. (some good communicators do) Normally it is just truism-suggestion, sometimes they can be linked but most people don't realise they are doing it so just use single sentences.

Another one could be

'You know where Johnny is? Can you call him for tea'

In sales

'Look at this phone; it meets all of your needs'

'You look like someone that likes making good decisions; this is the TV for you'

'You want the Big Mac Meal, and you're going large with that' (Question said as a statement)

On TV

'The question is shown on the screen; phone in if you know the answer'

'It's the end of the show; enter this competition to win £5000'

Contingent Suggestions

Some examples of contingent suggestions you may hear in everyday situations:

You don't have to brush your teeth until you're about to go to bed

When you go to the shop remember to get some milk

Wash your hands before you eat dinner

I'll read you a story when you're in bed

You can have chocolate fudge cake after you have finished your dinner

Contingent suggestions make one part of a sentence contingent on the other. The way to word them is to have the contingent part an unconscious process. If it is unconscious the client can't say 'no' when the behaviour it is linked to is true and happening

'As you blink in that special way you can become more absorbed'

'As you breathe out you can relax deeper'

'As you look at me, you can also be aware of certain thoughts that come to mind...as you become aware of those thoughts you can wonder what is happening in those hands...as you wonder what is happening in those hands you can notice that one hand feels different from the other...'

All pacing and leading and all starting with a truism. The contingent parts are all out of conscious control. Becoming absorbed, relaxing deeper, having thoughts, wondering what is happening in the hands, hands feeling different from each other. Nominalisations obviously help here with the leading parts.

When are people in a trance naturally?

I would say when people lose car keys that are right under their nose they are in a trance state. In the same way that when you get catalepsy in a cinema you are in a trance, when you forget a name at a party that you know you know and the harder you try to recall the name the more elusive it becomes, you are in a trance state.

I think it all comes down to how you are defining trance. If you get into a state of uncontrollable laughter you are in a trance state, same with problems like depression, smoking, anxiety, etc they all involve going into a trance and at the time you are in that state you see the world through that trance. Change trances and you see the world differently.

With the key example assuming they are in view and somewhere you have looked and not seen them often people find it is when they need the keys, they focus all their attention on 'where are the keys', then when they can't find them focus on how they can't find them and begin to narrow their attention on the issue of keys missing and as heightened emotion also induces deep trances (like phobias, fetishes etc) they are now getting emotional (stressed, anxious etc) because they can't find the keys so the trance gets deeper and more powerful and it cycles round as a self-fulfilling prophecy. The answer (as with the name example etc) is to stop, and think about something completely different to break the cycle.

One thing that gave me confidence at inducing a trance was seeing that everyone was going in and out of trances all by themselves all the time (also see Rossi; The 20 minute break - a book about Ultradian Rhythms). Leisure activities induce trance, reading, listening to music, daydreaming. Most trances people go into are self-induced and most people wouldn't notice that someone is in a trance because they wouldn't be looking for it. Driving involves trance (sometimes deeper than other times) you have to do many tasks simultaneously without thought, same as tying shoe laces, doodling, all automatic behaviours involve mini trances (handshakes, etc) A hypnotherapist can interrupt these trances and extend them and become a focal part of them to take control of the trance.

Binds, Double Binds & Implication

If you use things like a double bind you presuppose one direction whilst they think they are always making the choices which you are then responding to.

I remember watching Erickson say to a client 'look at my dog, what breed do you think it is?' The client wasn't asked if a dog was there only the breed.

If you are stating truisms people are not necessarily going to notice you are using a technique, and they can't really find holes in it. Also if you ask questions with implied responses but not actually asking for verbal responses or questions they would seem stupid to answer no to (like 'so your name's Steph?', or clarifying age etc...) it can just seem like you are clarifying rather than trying some technique.

Implied responses could be

'So you're sitting in that chair, and you can notice me and hear my voice and you don't expect to go into a trance yet' (Four implied agreements that don't ask for a response so are unlikely to face resistance etc...)

Post Hypnotic Suggestions (PHS)

Whenever anyone carries out a post hypnotic suggestion they go back into a trance like they were in when the suggestion was created to carry out the suggestion. If this trance is interrupted before finishing the PHS then you can expand on it and utilise it.

If someone has constant pain and it is ok to remove or alter the pain then you may want a client to hallucinate numbness or a different sensation for a long period of time (perhaps with conditions that if the signal is required it will come through)

You may not want to tell them it is only an hallucination and you may want them to be stuck in the hallucination to the extent that the pain control lasts a while. And that a trigger like opening the eyes in the morning could be used as a PHS for the hallucination to begin each morning.

At the end of therapy you would end everything you don't want them leaving therapy with so that they are completely reoriented back to 'reality' before going home. You may say 'you can wake up totally and completely' or 'wake up all over'

Nominalisations

Another area with therapeutic nominalisations is building your own context through the links between the nominalisations. If Development was used with talk of business the meaning of development to the listener is more likely to be in the context of business and it could be good or bad. If development was used in the context of 'what is happening now' then it is more likely to bring up meaning in this context.

The context the nominalisation is given in effects the meaning of the nominalisation.

'New developments are happening in the business, there will be organisational changes taking place'

'New developments are happening inside your mind, and you can wonder how those organisational changes and improvements will take place'

Fractionation

Fractionation is where you put someone in a trance and out of trance repeatedly. Each new trance induction deepens the trance. Erickson noticed each time he hypnotised patients they would go deeper than the previous time. He would spend weeks (sometimes many months) hypnotising clients. Then he wondered if he did the same number of inductions in a session rather than over many sessions would it have the same effect and he found out it did. To make it more effective it is useful to not bring the person 'out' between inductions. Just distract them (if they have their eyes open) or ask them to open their eyes (and not ask them to wake up etc) and talk and then do another induction (all can be indirect or it could be overt and asking them to 'close their eyes')

Inducing trance with music

You can induce trance with music. Many cultures have used music and no words to induce altered states (trance states). Many tribal cultures use flickering lights of the flames of a fire and drum beats to induce the trance state. Sometimes this beating can be fast, other times it could be slow. The trance states are different depending on the speed of the beat but it can all be used. I often use drum beats and other sounds to induce trance on my mp3 tracks and use binaural beats (different frequency of beats to each ear to create an illusion of a beat out of the difference between the two beats that often leads to the frequency of the brain matching these beats)

This all happens in music, especially modern music that can be listened to in stereo (or 5.1 Surround Sound) that allows musicians to create tracks that allow for deeper absorption from the listener. So any music can probably be used to induce a trance. I've known a number of teenagers I've worked with that would use Eminem or a gangsta rap group...If anything (especially with this age group) it shows respect to them that what they like and are into isn't being dismissed. I feel the times I've used these musicians/types of music that if I dismissed it and used something else it may have appeared disrespectful. I appear to show a keen interest if I ask for more detail and use it. You don't even have to know much about what you are discussing because you can be vague with your language and use nominalisations so it sounds meaningful to them.

Richard Bandlers Neurosonics hypnosis tracks use Blues and jazz to help induce a trance...I have used marching as a trance induction (ex-soldier), I have used dance music (trance music) and even rollercoaster's, through to diving, hang gliding, etc...Almost anything can be used...

Every day Trance Phenomena

what is commonly thought of as deep trance phenomena can occur in a lighter trance, like catalepsy occurring when watching a film in the cinema, or someone hallucinating that they thought they heard someone say their name, or hallucinating that they saw something out of the corner of their eye after watching a scary film (obviously these examples of naturally occurring times and so aren't as dramatic as when under hypnosis, but they are still occurring). I have had arm catalepsy in people in seconds while they are in a light trance. Noticing multiple deep trance phenomena or signs is more likely to mean deep trance is present than just noticing one (same as trance indicators like REM, fluttering eye lids, etc).

Rapid inductions

To do them the main thing is confidence. It is about interrupting a pattern or causing confusion. I rarely use rapid inductions. I rarely in private practice use anything that resembles a formal induction. At best I often ask people to close their eyes and I begin to talk to them and observe them closely, slowing down my breathing and speech, lowering my voice and tonality, and I mention talking to the unconscious part of them and just do it.

Really the inductions I do don't exist. For example if I am treating a phobia I just say something like 'OK just close your eyes a minute and we'll try something' then I go into the technique and make sure that all of my non-verbal behaviour is implying trance induction and deepening while they focus on the words and following any instructions (which is also trance inducing).

Watching Bandler, or some of Milton Erickson's footage can help. Also reading 'The deep trance training manual' by Igor Ledochowski and 'Training Trances' by Overdurf & Silverthorn is useful. These have info about rapid inductions and handshake inductions.

I do handshake inductions similar to how Erickson did them as they are less dramatic, or I will tell someone that 'in a moment I'm going to reach over and lift up your right arm...and I'm not going to tell you to put it down (implying levitation)...any faster than...your unconscious (embedded command)...allows you to go deeply and comfortably into a trance while the conscious part of you can drift off and think of pleasant memories or hopes and dreams...etc...' Then I reach over lift the arm very gently so I am hardly touching it and once it is levitating I say 'that's it' or 'that's right', as it starts lowering I suggest it can take its time etc... The person will already be in a trance, and as the arm lowers they will go deeper.

Or I do the fingers coming together induction. This induction is rapid and the client can then recreate it (if this is suggested) as a self-hypnosis induction in the future...

Observation Skills

When I first learnt about congruency between conscious and unconscious messages I wanted to know how I could practice this and refine it as a skill.

The best way I have found is to watch people, watch them in pubs, clubs, restaurants, anywhere where you get to observe people interacting. By doing this you can listen to conversations at the same time as objectively watching non-verbal behaviour. Another place to watch this is on reality TV shows like Big Brother and on programmes like 'the Jeremy Kyle show'. I used to record one of these shows a week and watch interactions and see what I could figure out about people based on mismatching communication. With programmes like Big Brother you can test your ideas about your observations over a period of time.

You can watch people talking and look for patterns. Doing this you don't get to ask the questions but you can pay your full attention because you aren't involved. Anyone that has knowledge of magic and watches a magician knows that if the magician is captivating enough you miss what they do even though you know it happened right under your nose. This is the same when starting out doing therapy, you know lots of stuff but miss it when you are in a real situation because you have too much to take in.

As you watch people you may work by initially just getting a sense of something or you may actively look for patterns that you could tell someone else (like change in facial colour, change in lips, body posture, eye contact, etc)

The best way to learn to recognise minimal cues is to focus on one at a time while you learn.

What you do with the observations depends on what you are observing for (it could be to look for congruence, or it could be for a specific state, etc) If it is for a state then you can suggest back the minimal cues, so if you wanted to induce a deep trance comment on the minimal cues (overtly or indirectly) each time you see a trance based minimal cue. You could link it to going deeper for example by saying 'as you continue to blink in that special way you can drift deeper.' Or 'That's Right' (said on each blink or sign of ideo-motor movement etc)

The easiest way of noticing minimal cues is to be in a trance, letting your unconscious notice for you.

To switch the trance focus (from internal to external or external to internal) you can start by matching the experience then guiding it to where it is wanted.

'You can be aware of the ticking clock, of the traffic outside, of the sound of my breathing AND you can notice what those hands feel like resting on your lap WHILE you wonder what will happen next...and BEFORE you discover what will happen next you can notice which hand feels the heaviest and wonder which one will lift...' (Getting more internal)

To do this the other way reverse the process and match on-going internal experience then you can ask them to remain in this state while they open their eyes and pay their full attention honestly and completely (a statement they should take literally) to ... (whatever the external thing is - reading, practicing an instrument etc)

With leisure activities you can have an external focus activity and guide it internally (even by saying 'I sense you can feel some of that now').

Six-Step Reframing

I remember Richard Bandler saying he created the six step reframing technique (find out more in NLP literature) as a way of getting people to do hypnosis (when he couldn't use terms like hypnosis) and talk with the clients unconscious (or have the client talk with their own unconscious mind) to allow the unconscious to hold multiple views (positive intention of the problem, how to find a better way of meeting that intention - like smoking to relax, self-hypnosis to relax - and to allow the unconscious mind to take control of integrating these ideas) He has said he doesn't use this technique as a technique now because there isn't such stigma about using hypnosis and just doing this directly like lifting an arm and saying not to lower any faster than you make these changes (while the person is in trance). The six step reframing was designed as a trance induction and therapy script all in one technique, 'go inside, thank your unconscious for its solution (the problem), ask the unconscious to find a variety of better options that fulfil the same need or requirements, check this is OK (look to the future etc) - EG - Smoking to relax, deep breathing to relax or self-hypnosis to relax or time management to relax etc, healthier and fitter in the future and some positive side effects may be happier at work or home etc) then give a signal when this is done (arm lowering, finger signal, feeling in the body, eyes opening etc)

Cause, Symptoms, Problem Perception & Solutions

Having cause and symptom both different views of the same thing is like a fish's view of an iceberg and a bird's eye view of the iceberg. The therapist and client may only be aware of the bird's eye view, by knowing that what you can observe is part of what is hidden not something different from what is hidden, you can look at what is keeping the top of the iceberg in place.

It maybe that the cause no longer exists but something is still keeping the iceberg tip in place, this could be an unconscious belief that the habit is still required as it served a purpose, which would mean there is still a linked cause (the belief it is needed - like a form of self-therapy, the cause being the therapy, needed or not - like insomnia because 40 years ago light sleeping was required and the pattern has stuck, or smoking to manage stress etc).

The similarities with particle physics are useful ideas and interesting to wonder how truly linked these things are. In quantum theory something is in every possible outcome until it is observed, then it settles on one outcome...the observation create the outcome. The position of the observer and what is observed also affects the outcome. This is the same in therapy, observation affects the perceived problem.

Also creating a change in one place creates an instant change in all related areas. So if someone that has an acne problem forgets they have and forgets to pay attention to it, it disappears quickly. Or if someone stops worrying, they stop being depressed, if someone tries to stay awake instead of doing their usual behaviour of trying to fall asleep they fall asleep. So changing one part of a pattern alters the whole pattern, leading to a different outcome.

This was a large part of Erickson's work. He wouldn't know what was going to happen, only where he could help make a change. He would trust the client to make it positively and in context with finding a solution to their problem.

For example when he knew a man could move one of his fingers, he got the man doing this, he didn't know exactly what the outcome would be only that when the man noticed he could move one finger and the finger next to it moves a little he'll know that means he can move a second finger, which moves a third finger slightly etc...

On Sunday I was in a car getting a lift back to Bognor Regis from Roehampton...It was raining quite heavy...we took a wrong turning...The driver asked if we should perhaps get out the satnav or hope we find our way or if we have got lost for a reason we don't yet know?

I told a story about a farmer that had a prize stallion that escaped from his field...the villagers say 'that was unlucky' the farmer says maybe, the stallion comes back with some healthy wild horses...the villagers say 'that was lucky' the farmer says maybe, the farmer's son gets on one of the new horses and gets thrown off breaking his leg...the villagers say 'that was unlucky' the farmer says maybe, the army come to town to recruit soldiers and the farmer's son can't be recruited because of his broken leg...the villagers say 'that was lucky' and the farmer says maybe...

A couple of minutes after telling this story when we were about 2 seconds from a petrol station (literally) the engine cut out and we drifted comfortably onto the forecourt in front of the correct pump to try filling up the car (it didn't need more petrol), we called the RAC and it turned out there was an RAC van less than 10 minutes away from where we were...no -one needed to use the pump we were at and the RAC man knew what the problem was and how to sort it out quickly and we got back to Bognor fine.

Had we not taken the wrong turning and just gone with it, we wouldn't have been in the right place at the right time...instead we would have broken down on a motorway in the rain and dark rather than under cover in a service station with a shop and just down the road from an RAC van.

The unconscious is very good at working in ideas, concepts and metaphors. For example I have seen a video of Erickson lift up a resistant clients arm palm facing them then putting it down again, then getting them lifting their own arm to the same position and putting it down again (to take down their own barrier between them - that is where the arm was placed)

Setting tasks does a similar thing like the example of Erickson having a client hunt for two identical blades of grass, or giving out African violets.

You only need to ignite the change not do it for the client.

Time Distortion

I once hypnotised someone to experience a whole lifetime in a world they created in the space of 10 minutes...They spent hours recounting what they got up to over their lifetime...

It happens when you dream, if you get asked about it straight away you can recall loads of information about the dream and can claim it was hours even if you know you only nodded off for a few minutes

Anaesthesia using time distortion is like taking different drugs. Some drugs may work on just the discomfort, other drugs may make the whole area numb, other drugs may make you unconscious (so having no awareness of pain). Pain is made worse than it needs to be because you do remember what the pain felt like, you think about how painful it will be in the future (both of which will bring back to some level the feelings associated with the pain, just like a leisure trance induction brings back to some extent the feelings associated with the activity) and you have the pain you are currently experiencing.

You could make the area of the pain numb (for example a pain in the wrist could have the WHOLE wrist made numb) or you could make JUST the pain numb (so that if you touched the wrist you would feel the touch because only the pain is numbed), you could give amnesia for past pain and have the person forget they may have future pain (reducing the pain by 2/3), you could change the sensation of the pain (so that it feels like a warmth or tingling etc - pain is a signal saying 'look after me' so sometimes getting rid of it completely will not work because it is important that you have some awareness of it) or you could speed up the passing of the pain (again respecting that it is there as a signal rather than just making it go away). Technically if the pain (headache for example) would have been there for an hour and it is gone in one minute then what is happening for the other 59 minutes? It must be anaesthetised because it isn't there when it would have been.

In the past I have asked people to experience entire artificial life times, almost like giving them the chance to live through a fantasy (like in the land of lord or the rings, or star wars, or on the star ship enterprise) and to take a minute of my time to do this. They experience the highs and lows of that life (I add bits to how I suggest it so that they don't have anything like being killed in their 'dream' and clarify what they want). It has been really interesting talking for hours with people when they come out of a trance and tell you their WHOLE life story, all the adventures etc as they grew old. People have told about living to 70, 80, 90 etc and recall memories from all of it. So I think the mind can distort time to an incredible range effortlessly.

Regarding Time Slowing Down, when I worked in incredibly violent and aggressive situations time would slow down to what seemed like a snail's pace. I would become so aware that for a five minute incident I could write pages of information about every little detail (in my job at the time I could write the events of a 24hr period in a single page and think I had covered everything). The same thing occurred when I was ran over, time slowed right down, all I felt was peace and relaxation. I was so aware of everything around me. Time changing like this in emergencies is a natural survival response to give you more chance of effectively managing the situation (if you are attacked and go into this state you are hyper aware so time appears to run slowly).

This type of time distortion is different to comparing spaces between events, which is why as you get older time appears to speed up (a year for a ten year old is 10% of his life, a year for a hundred year old is 1% of his life) so time appears to get faster. Also amnesia between events makes time seem faster.

Generally time distortion will usually happen whether it is asked for or not. The best way to elicit time distortion is through stories, metaphors and examples of naturally occurring time distortion. Creating a situation conducive of time distortion, like getting the client very engrossed or focused on one thing so they exclude all else, or language patterns. It could be directly suggested with some people but that doesn't work with many others. The use of language patterns to induce time distortion fascinates me, you could say something like 'in hypnotic time you can discover a whole hour seems like a minute, as in waking time a whole minute can stretch to an hour' or another variation maybe 'You can be curious to discover whether an hour of your time will seem like a minute of waking time or whether a minute of my time will seem like an hour of hypnotic time?'

Analogue Marking

With the issue of analogue marking or embedded commands, I use it all the time throughout whole sessions. It is like allowing your communication to be multi layered. The conscious mind will be phasing in and out.

I think it is important to be continually allowing communication to both the conscious part and the unconscious part. Analogue marking allows the conscious part to listen into an apparent conversation whilst the unconscious part is aware of the marked out sections.

Because the unconscious part notices these sections and the conscious part doesn't the conscious and unconscious receive two different messages. I often do this telling stories/metaphors etc that the conscious mind just listens to while the unconscious part responds to the patterns in what I am saying and also to any sections that get marked out (a form of analogue marking is embedded commands/suggestions).

The 'my friend john' technique is a good example of this used in trance induction. It also happens in everyday life, you get people that say

'I told him...I'm really annoyed at the lack of respect you show me...' As you hear someone talking directly at you like this it can feel like it is aimed at you, it creates feelings in your body as it affects you on a deeper level even though logically and consciously you know they are not talking about you.

If ideas and suggestions are given indirectly (via analogue marking for example) then the conscious mind is highly unlikely to notice so it will only be received unconsciously, if the suggestions were given directly then the conscious mind may become aware and may in the future sabotage the work because it remembers bits and pieces.

When I want to educate someone in therapy and feel that they probably don't see that they don't know what they don't know I do it indirectly. Often by telling them I'm not going to tell them. because I am there as a therapist and they have come to me for help there is a high chance that they expect me to know what I'm talking about so if I don't give them reason to challenge me then often what I say gets accepted.

As an example in smoking, some people think they know the risks of smoking but don't really, they only know the common few things that get plastered over the media. I want them to have an understanding of some of the other issues but I don't want to lecture them or to have them defend why it won't be them etc (I don't always feel this is necessary, it is client dependant).

I will often say 'I know you know all the effects of smoking, so I don't need to tell you that 50% of all smokers die of a smoking related illness'...then I tell them what I said I wasn't going to tell them.

Rapport

Over the years I have found that many people misunderstand what they should be doing when building rapport and matching or mirroring body language etc. People seem to think they should be copying but this really annoys people. The idea is to join the person in their reality respectfully. Look for patterns. If they do a specific gesture when saying or talking about a specific thing (like making a brushing motion when talking about getting rid of a habit) then if you talk about the same you can make the same motion. You don't just do it because they did.

I match breathing and often match heart rate with the movement of my head or a finger or my foot, often match blinking (either with my blinking or with a finger etc), I match general body posture and certain words and phrases used. I don't match things that would be unnatural for me (I had a client that had a bad arm and she kept it in an awkward position. I could copy it but it would have been uncomfortable and would have looked like I was taking the piss).

If you are matching how someone is sitting and they change positions, don't immediately change how you are sitting, wait until a time when it would be natural for you to decide to change positions.

The idea is to respond in the same way they would, not to copy them. I pace then test then lead people into trance without placing any importance on the words I am saying just by using breathing, body posture, heart rate, blinking etc so that when they are responding to me I gradually put myself in trance and they follow.

Working with Problems

You should allow the client enough time to talk about the depression (or their problem as they see it) it would seem very disrespectful if it was ignored or brushed under the carpet. Just like you can use questions and asking for more detail to induce a positive trance, you can do the same to induce a negative trance, so if you ask them questions about all of their problems and make your focus on their problems they'll give you the answers whilst at the same time they'll often begin to deepen their depression because your eliciting proof that they have had a lifetime of depression (which backs up the depressive thinking style).

If you listen to their story and focus in on the islands of hope and resiliency whilst at the same time acknowledging 'that must have been a difficult period in your life, no wonder you got depressed, I'd be surprised if anyone could go through a situation like that and not need to take some mental time out to get your head round it before being able to move on' (or whatever wording/comment/etc would feel right for the client in front of you)

It's not that the therapist shouldn't talk about the presenting problem (depression), It's that I don't believe the therapist needs to talk about all the depressing things that have happened to the person in their life on top of that problem (as some therapists I've had to observe have done, and many clients I have had have been the result of feeling suicidal after seeing counsellors that have done this leading to them believing they were justified to think their life was a failure and worthless) I like people to leave a session optimistic and with a sense of hope and expectancy for positive change

Smoking is a trance state, the craving focuses attention and negates everything else, I've known people that have said when a craving happens they would kill for a cigarette, people have said to me that they turn into a different person if they try to ignore the craving, they 'see the world through the filer of the craving', when they smoke the process happens automatically, they don't think about what they are doing.

Often in therapy I ask them if they enjoy every cigarette? And if they remember each cigarette? The answers are always that they just need a cigarette and find themselves smoking. If they make roll ups then they do this whilst talking, etc with no conscious thought, they get on with chatting as they get a cigarette out and light it, they instinctively light up a cigarette when they have a coffee, or when they answer the phone, or when they get out of a no smoking area (or whatever it happens to be for them).

What I do often is (especially with people that don't want to quit but have been told they are seeing me to quit) is make it so that it becomes so conscious as a process they don't enjoy it. I don't do this by working with the conscious part of them directly, so I don't tell them what I want them to do. I do things like scrambling their usual pattern of how their smoking habit works, by having (instinctively) the negatives coming to mind when they think about smoking or go to smoke, and positives coming to mind when they think about being healthy and when they decline offers of cigarettes etc. Because this is coming to mind for them they are very conscious of the act of smoking (not because I have asked them to be but because it is what comes through from their mind when they try). I, at the same time presuppose when they will finally decide 'enough is enough' for themselves and offer choices about that date, etc...

I also often make it conscious by suggesting tasks, for example a man saw me (referred by the NHS stop smoking clinic). The first thing he said was 'I will tell you the same as I told the nurse at the clinic, I will have you smoking 50 a day before you would be able to stop me smoking 50 a day' he knew why he should stop and was not there because he had any intention to stop.

One of his regular patterns was to smoke a specific number of cigarettes on his set journey (like to the supermarket) he had a reason for doing this. For example to the supermarket he had to smoke 6 cigarettes, he would have 4 on the journey (15 minute journey) and 2 in the car park. This was because once he's in the shop he can't smoke until he comes out. He agreed he can go 15 minutes without a cigarette (does longer in the supermarket) he agreed he won't cut down or stop, he also agreed that he was willing to try my idea of really making the most of those cigarettes and smoking them all when he is in the car park or when he has arrived at his destination, if he isn't driving he can really focus on enjoying those cigarettes.

I told him but I don't want you to cheat, you must smoke all the cigarettes that you normally would have smoked. He found it such a chore to be sat consciously smoking each cigarette when he was at his destination and felt he was wasting his time because he wanted to be doing what he was there for not just sitting in his car smoking. He saw me a week later and said he had tried to do as I asked but found himself cheating sometimes and not smoking as many as he was supposed to be smoking.

He then cancelled the next appointment and quit on his own without help so that he could take credit for it (my assumption was that he probably felt he would lose face if he quit because of a therapist when he told all therapist none of them could stop him) Months later I saw him and he had lost weight, not smoked and looked healthier and was proud he did it on his own.

All psychological problems involve a trance element...this is one reason why hypnotherapy is so useful for treating them. Each trances has different characteristics, a depression trance offers a specific view of the world that is different to a happiness trance for example.

Addictions and eating disorders and OCD, phobias, PTSD all involve a trance element. An easy way to hypnotise a smoker for example is to ask them to talk about how they smoke, what feelings are associated with it etc. Even the trances therapists guide people into can be different. It is possible to create 'designer trances' where you create a desired trance then use fractionation to add a different trance onto that one and have different combinations of trance.

I saw someone the other day that had one type of trance experience with me a week earlier and told me what he liked and what could make the previous experience better so I helped him experience this new type of trance. To get it I had to do different things. Different rhythms and frequencies create different trance states. I made an altered states mp3 that involved using drum beats (and different drums and spatial locations when listened to in stereo) to create an interesting trance state.

Obviously it is beneficial to use something enjoyable and positive as the induction but it is also useful to know that the problem is a trance state and a unique one at that and so any alteration made to that trance state will change the resulting trance experience or break it completely and make the unconscious processes leading to the trance conscious. (Like making a smoker aware of every step of smoking by altering an element of it so that it is no longer an unconscious habit but more a conscious chore or something that suddenly they feel self-conscious about because they are now aware they are doing it)

In my experience with people with over drinking problems I've found that often it is their medication, they drink to self-medicate against something. It could be that they get too stressed and so drink to forget what is causing them stress, or they have a drink and know when they go home they will be late and will end up in an argument so they drink more and don't go home yet to face the music, which leads to more problems when they do go home. Very often they desperately want to quit but as long as they don't know of a better more productive way of dealing with what they are using the drink to deal with then the pattern continues.

People may drink when they socialise and believe that they won't be able to socialise if they stop drinking, or when they drink they get attention and looked after the next time they are sober, etc... Normally in my experience if there is a secondary gain then it stands out.

I have often asked people 'so what do you get out of drinking/smoking etc?'

Or asking about relationships (friends and family and work colleagues), normally as you take a history and cover all aspects of their life surrounding the problem you observe certain things that could be secondary gains (like a wife looking after the husband when he has sobered up, or special treatment from work colleagues)

I've worked with children through to adults with mental health issues and I don't mind if they are talkative or not talkative at all or highly educated or don't understand most things most people would understand. I find that if I utilise whatever they present and if I talk in a way they can understand. I feel that the therapist should be flexible and adaptable and change their approach, words and actions to suit the client. Sometimes I may use child type metaphors like stories and fairy tales other times more grown up examples perhaps of daily situations. I wouldn't want them to shut off so to keep them happy I want them to feel I'm understanding and respecting what they know. I may not use so many complex language patterns, not because I don't think their unconscious can understand, but because it may sound complex and break rapport. I'm more likely to stick to embedded commands. The same for gathering information about the patients' history I would want to do this in a way that suits the person I'm working with.

The patterns I look for are more the structure of the problem. The content of the problem is often very important to the client; this is what they have often come in with and what they feel the therapist needs to hear. It is important to respect the client and to acknowledge their feelings and circumstances and to feed back in an accepting manner that you listened and that times perhaps were not easy. If there are exceptions these can be fed back and islands of hope and success can be feedback. Also if they mention positive up and coming things these can be feedback. It's more a case of positively directing the clients' attention whilst respecting what is also important to them and showing this respect. You will only be feeding back truisms (statements of fact) so anything you say is unlikely to be challenged or seem disrespectful. With parents they often say they reached the end of their tether and just walked away and then the child calmed down. I often feedback when summarising '...and how did you know that walking away was the right option for getting Joe Blogs to stop shouting and swearing?

As all of the summarising is truisms the yes set created will often aid continued rapport and often your summarising makes them feel better and more hopeful (which helps rapport to)

I think often an outcome a therapist comes up with is not the therapist's outcome but an observed outcome that the client hadn't made the connection to. I believe that therapists should be humble enough to respect a client's view on what they want treated and what they currently don't want treated. I knew someone that wanted to stop being depressed and they smoked. I helped them stop being depressed but didn't try to MAKE them stop smoking. Another therapist was horrified that I sent them on their way 'imperfect'; their training had been that a client stays in therapy until they have no more problems. My question was by whose definition and standards? And what is a template of a normal problem free person that everyone should leave therapy fitting? This depressed person had two sessions and was happy with the results (and actually cut down on smoking as a side effect of managing stress more effectively, and managing boredom more effectively)

Some of the most important information is gathered from asking about when the problem doesn't occur, especially from when the problem doesn't occur but was expected to...(for example; if a smoker HAS to smoke every hour and then goes on a 7hr flight to Dubai and doesn't even think about wanting a cigarette).

This information can be talked about and expanded. Also for example if someone is ALWAYS depressed and you ask them about the times they feel less depressed (if asking about times when they weren't depressed is too much of a leap) you can find patterns associated with the non-depression, or happier times (at this point they may not accept that they were happy or having fun at the time but being less depressed is a start).

Asking about exceptions builds hope and expectancy as you guide the client's attention to solutions and problem free times. If they think they always have the problem (like pain or depression etc) and now they have examples of times without the problem it begins to break it down.

I remember hearing about a lawyer doing this...Someone took the stand with pain, they were trying to get compensation for an injury, and the interaction went something like this;

Lawyer 'where is the pain worst?'

Person 'in my right arm'

'Where in the right arm is the pain worse?' 'In the lower part of my arm' 'so the top part of your arm is more comfortable?' 'Yes' where in that lower part of your arm has the least discomfort?' 'My elbow and down my forearm' 'so your elbow and forearm are more comfortable, is the feeling more in your wrist or your hand?' 'More in my hand' 'so your wrist is more comfortable than your hand?' 'Yes' 'which parts of your hand feels most comfortable, your palm or your fingers?' 'my palm' 'your palm feels more comfortable than your fingers' 'yes' 'which of your fingers feels most comfortable?' (And the questioning continued down to the thumb) '...and when does the discomfort affect you least, day times or night time?' 'Day time is better than at night' 'does it bother you while you sleep?' 'no' 'do you feel more comfortable when you wake up or when you fall asleep' 'when I wake up' (at this point recapping is done) 'so you've got some pain, its better during the day, it doesn't really bother you while you sleep or when you wake up, and it is worse in your right thumb just when you are falling asleep and more comfortable in the rest of your hand and arm...'

The person taking the stand apparently was more pain free and appeared happier but didn't get much compensation...

Every day I use my mind to control discomfort. Time distortion and perception fascinates me. The uses are varied. When you are in pain you are naturally in a trance so a trained Doctor could utilise this by building a yes set around the feeling then wondering curiously how long it will last and what it will feel like as it disappears and whether it will disappear from the arm up or the neck down, or from the centre out dispersing or maybe round the edges first or maybe in a random fashion... It is interesting to offer a client more time in a short space of time to do some work by talking about natural times time distortion happens. Like when you are doing something exciting and time flies by, or when you read a good book and notice you've nearly finished and been sat there for hours but it seems like only a few minutes, the same when watching a good film. And I'm sure you can think of more?

Another use for summarising is reframing. It gives you a chance to offer a slightly altered interpretation that is more beneficial. Like changing pain to discomfort (which hypnotically has the added bonus of being 70% made up of comfort). You can phrase the feedback as 'correct me if I'm wrong....'

I really do value and appreciate the use of questions. If you ask the right questions, you guide the client through a more productive route on their map of reality. It is like they have been walking through a dark and misty moor unable to see the safe route through so they keep getting stuck. If you ask the right questions it's like you're helping to guide them along the safe path to where they want to go.

Even though they may not realise it your questions can focus them on what they should focus on. If you have a depressed client and you ask to be told about all the bad things that have happened in their life they will tell you and may get more depressed. If you ask them for the exceptions, for when things have gone well, etc... You are more likely to focus them positively and give them hope, same as if you make their problem seem more normal.

The same when setting tasks, you can ask them to notice what happens between now and the next session that they would like to continue to have happen.

I work with many clients; the questioning is a very therapeutic tool as it gathers information that is useful directing a client around their map in a way they may not have taken before. For example; I work a lot with parents that have children with behavioural problems. Often I meet the parents or parent and they have 'rehearsed' what they think they need to say to me and what they think I need to know.

I often start a session talking about irrelevant seeming things like how 'it's a nice day for a change' (one of my favourite ambiguous comments to start a session), I try to notice something about them that I can talk with them about, I ask them what has been happening since they made the appointment that they would like to continue to happen, or I may ask them what has been going well, etc...I'll try to begin to focus on successes. I let them tell their story for a while, but often feed key bits back.

For example; Parents often say when describing the problem that they reach a point where they have had enough and they just walk away. I often ask what happens next? The response is usually that they calm down. I then often comment 'How did you know that to make him calm down and to end the situation you walk away?' I don't mind whether they accept my interpretation or not at this point, I just want to get across an alternative view for their behaviour as a success not a failure.

So I think skilful questioning and responding to responses is therapeutic. On average in an hour session I probably spend about 15-20 minutes gathering information/history about the problem, the process, how the problem starts and ends, a few examples (so that I can hear them describing it), the context, whether they have solved the problem in the past (For example; perhaps they want to quit smoking and they did for 5 years 10 years ago, then I can find out what is stopping them doing the same again and what made them restart), etc... but with some people they have a need to talk for longer and feel listened to and other people give what you probably need really quickly (I help someone with a phobia once in less than 10 minutes because the situation only allowed this and they needed the phobia removed, I gave suggestions at the end of the work to the unconscious to use day dreams and night dreams to make any other necessary adjustments to make the change comfortable and lasting whilst maintaining any necessary learning that the old incident had taught)

What I have observed is that it isn't always necessary to know where the problem came from to treat it, the cause may have long gone, I often give a few metaphors/stories about change and work with what information I have, often if the person can really get a sense of the future without the problem (if they struggle to accept the possibility of change and describing a future without the problem I sometimes say 'what would it be like if...' or 'Imagine you wake up tomorrow and what bought you here today has stopped, how would you know...etc') then they rarely need to go back and find why it was there or deal with any past issue, it may have burnt itself out or passed its sell by date and no longer be relevant.

What I often find is that if the first session doesn't get significant results then I may have to dig a little deeper or even use ideo-motor responses and ask for the unconscious to work things out and signal yes when it is done.

Looking for Patterns

There are many examples, people shaking their head while they say yes (or nodding and saying no) this movement will often be slow and fairly minimal, crossing arms (or legs) whilst trying to appear like they don't have a problem with what you are saying (or pretending to agree), people acting interested with their feet pointing at the door (or someone they would rather be talking with), micro movements (brief glimmers from the unconscious to what has just been said or happened moments before the conscious mind responds) like slight scowls followed by a smile or part of the face giving a response like eye brows lifting like fear or startle as the person tries to smile and not look surprised. Metaphorical behaviours seem to be very common, people having an aching neck or digging heals in, or using hand gestures to push a problem away or to move something closer or to put things in place.

In my experience it is about looking for patterns. For example someone may cross their arms because they are cold or because they disagree with something you have said. My advice would be to be observant. If you think you have noticed a pattern (someone crosses their arms while you say something) then change subject (they may then uncross their arms or may not if they are cold) then a few minutes later go back to the subject again and see if they carry out the behaviour again. Same with rubbing the neck. It could be a genuine pain in the neck or it could be a husband being a pain in the neck. If they rub their neck when the husband is mentioned and then you change topic and come back to it and again they rub their neck it may well be a sign. I have known of many courses that teach about body language and non-verbal behaviour wrong (in my opinion) they say 'this means this' in a very rigid way. I had an interview where the interview panel all interpreted me shifting on my seat and crossing my arms as being very uncomfortable talking about my past (I had just been ran over and so had to cross my arms to support my bad arm and had to shift in my seat for the same reason. Over a prolonged interview it was inevitable (to me) that I would begin to get very fidgety)

Introduction to Part Five 'Therapy in Action – Performance Enhancement'

This is a transcripts of a therapy session that includes my analysis to show what is going on to create change and to explain a little of what I am doing.

In this session I am helping the client to improve their artistic abilities. I have looked at research into creating savant abilities in people and thought it may be possible to do this with hypnosis.

The session was an hour long and the only session required to help this client improve their artistic abilities and to still (about 5 years later at the time of writing this) maintain and develop that improvement further as time has gone on.

I use a 'D' for when I am speaking and a 'C' for the client. The analysis of what I am doing is 'cut in' to the session where I have felt it is useful to note techniques or language patterns I am using. Hopefully this session will give a good overview of how the techniques and learnings from this book can be applied during therapy helping to bridge the gap between theoretical learning and practical application of hypnotic and therapeutic skills.

Art Improvement Session

I ask the client to draw a picture of a horse in one minute and at the end will ask them to do the same again.

While this session was taking place I am matching the clients arm positions, breathing, leg positions and leaning in at about the same angle the client is leaning back. This all helps with rapport and building a deeper connection with the client and their model of reality.

Initiating Trance

D: (Looking down from the client) When was the last time you (looking up at the client) **went into a trance**

C: Mmm…I don't know…probably in mmm, I went into a trance…well…probably last week…Tuesday…Tuesday this week because I was pruning my bonsai tree

D: When was the last time you (looking at the client in the same way I did above) **went into a deep trance**… (looking down) do you remember (looking back up at the client) **what one is like**…

C: Yep

D: **You do**

C: Vaguely

How I often start out inducing a hypnotic trance is by getting the client to recall a previous trance experience. If a client says they have never been in a trance before then I ask them what they think it will be like or what they expect it to be like; or I ask them about everyday trance states like leisure activities.

All the **highlighted** words are embedded commands or suggestions. They are parts of the communication I am adding extra emphasis to by using a more hypnotic voice, defocusing my eyes and relaxing my facial muscles all to imply trance through modelling what I am expecting.

I imply that the client has got experiences of entering a trance and deep trance by using the term 'when'; I don't ask 'have you ever been in a trance?' In some cases I may ask 'have you' if I strongly suspect they may not understand or realise that they have been in a trance before. If I do this I would use the term 'before' at the end of the sentence as 'before' implies either before the one they are in now or the one they will be going into.

I very often use feeding back as a way of embedding suggestions or commands; for example when the client says 'yep' and I respond with the command 'you do'. I will often turn questions to statements with the tone of my voice. By feeding back what was just said you also begin to form a 'yes set' this is where you get yes responses that help to build rapport; displays understanding and makes it more difficult for the client to respond negatively to the work you do. If you get someone to say three or four yes's then give a suggestion you want them to follow they are more likely to do it than if you just gave them the suggestion on its own.

This client showed signs of not wanting to commit just yet by responding to my feeding back with the response 'vaguely' rather than responding by saying 'yes'. He had just answered my question by saying that he can remember what a deep trance is like but almost immediately became indecisive.

Seeding/Priming Using Metaphor

Sowing the seeds for amnesia and entering a trance when drawing

> D: (starting to tell a true story to seed what is to come in the session, as I do I look down) You know when I was a kid I used to like watching...I can't remember the name of the programme now...that Rolfe Harris (turn to face the client) **drawing**...(keep facing the client but move my head position) kids programme that was on CITV...
>
> C: Rolfe's Cartoon Club

D: Yeah something like that (I look down again) he used to have that like (drawing a circle with my arm and finger) (looking back at the client) **circular desk thing**

C: (Doing an impression of Rolfe Harris)

D: (looking down again) And Mmm I (looking back at the client) **got the book of that** (Looking down again) I don't know how and I don't know what happened to it…it's not like me to lose a book…mmm…but I (looking back at the client) **always wanted to draw the Rolfaroo** (drawing in the air again)

C: Ahh yeah

I now use a true story to begin to plant the idea of automatic drawing. I also want the client to have amnesia for much of the session so that later they don't try to review everything that I was doing. This specific client has considerable experience with using hypnosis so I want to make sure they don't analyse the session too much when they come out of a trance. I also want to tell a story that lays down the pattern of the session we are in now where I want the client to start off more conscious, then the unconscious can take over more. I also have chosen to use a story based on being a child of about eight years old because that way the client will begin to make associations unconsciously with when they were eight and this will cause a slight regression and will change the clients beliefs and opinions of what is possible; as an eight year old almost anything is possible.

I am still continuing to embed suggestions and ideas. While I am telling the story I am drawing in the air (like automatic drawing taking place) and putting emphasis on the words I am saying while I do this so that the client knows that these parts are important.

> D: And in the book (gesturing where the book is) it tells you how to draw it, but I could never draw it, I could never get it to look like the one Rolfe Harris (looking at client) **draws** (drawing in the air again)...(looking away) and I was 8 or something...didn't matter how hard I tried to draw it, it never looked like the ones Rolfe Harris draws...then one day I was in the lounge trying to **draw it** (drawing in the air)...kept failing, well not failing, I kept drawing it but not doing a good job...so I sort of gave up and still had the paper and still had the pen on me, and was watching something else, and then I thought 'yeah lets draw it again' (mimic going to draw) (a car alarm goes

off)…then I got distracted, by the TV…I can't remember what I was watching now but it distracted me and I suddenly thought 'cool I'll watch that'…didn't even think…It was one of those things where you know if you put an action movie on like Die Hard or something, and not a lot's happening so you're happy to read or do whatever, and then all of a sudden you hear some action happen and so you look up at the screen and think 'cool' and mmm so I looked up at the screen but my hand kept drawing…and I looked back down…and…like the cinema thing…you can be eating popcorn and something exciting happens and all of a sudden your hand freezes in front of your mouth because you're more concerned about (gesturing in front of me and looking in front of me – congruent – and gesturing holding a piece of popcorn in front of my mouth) what is going on on the TV and all of a sudden that finishes and you carry on putting the food in your mouth (gesture doing this)…as if…you know…**the interruption never happened**…and but what happened was my hand didn't stop, so when I was interrupted and I looked up, my hand continued to draw…and I looked down at a perfect Rolferoo and I was really proud except that it surprised me because I couldn't do it…(simulating trying to draw) never mind how hard I tried to draw it I couldn't draw it, and it took me by surprise because I did it without looking so I didn't know where anything was, so I shouldn't have been able to get the eye in the right place and I didn't know where the paper was…and people touch type and they can't see where all the keys are…they can get their fingers to the right keys…they can't see it at all…they can do it with their eyes closed and you know people play the piano the same…and people play the piano blind, like Stevie Wonder…so obviously it is possible for your body to know things that you don't know…like my granddad plays the piano without looking at the keyboard…and it is just one of those things…yes so Granddad can play the piano and not look at the keys…mmm could probably play in his sleep…and it's always fascinated me, I **can now**…I've never learnt to touch type, but I **can now** type to some extent without looking at the keys, as long as I don't consciously make the effort not to look at the keys, if I just glance up at the screen I can **keep my fingers going** (acting out typing without looking at my fingers) … If I consciously think 'Oh I've been typing without looking at my hands' all of a sudden I'm pressing all the wrong buttons trying to type…

As the story continues I talk about how trying hard can lead to failure, or at least not getting the success you want. This is important because trying is a conscious process. And generally the harder you try the harder something seems to become. In the same way that if you try to fall asleep you struggle and stay awake, and if you try to stay awake you fall asleep. What I want is for the client to just do, not try.

When a car alarm went off outside I decided to incorporate this into what I was saying so I spoke at that point of the distraction that gets your attention for a little while. I then continued to parallel the alarm by saying about an action film like Die Hard and hearing something happening that gets your attention. While I told this piece I suggested amnesia and also suggested it being a positive distraction by interpreting the distraction in my story as 'cool'. I then talk about the hand continuing to draw while I am distracted partly because I want this process to happen in the client when he draws but also because if the alarm keeps going off or goes off again I want the client to be still responding on an unconscious level regardless of external distractions.

I create frustration in the story so that the client would become more focused on what I was saying. I spoke about how my hand continued moving then I ended that piece at an 'and…' before changing to talk about being in a cinema. What this does is a part of the client stays focused on looking for the completion of the pattern/story; and while that is happening the client enters a slightly deeper trance giving me the opportunity while they are partially listening and partially focused on waiting for the end of the story to continue to embed the idea of amnesia that I have been embedding already and also metaphorically commenting that the car alarm has stopped now.

I then carry on telling the story to convey the message that while the conscious mind is occupied with one thing the unconscious mind can work on something else. I then give different examples of being able to do things without conscious attention. Each example is true and undeniable which helps to convey the message that what we are doing is also possible. I use these stories to convey the message of not paying conscious attention, being pleasantly surprised by what the unconscious has done, which also contains the message that the conscious won't be a part of the process.

Towards the end of the story I start to separate conscious and unconscious by talking leaning slightly in one direction when talking about the unconscious mind and the other direction when talking about the conscious mind. I say '…it is possible for your body to know things that you don't know'. As I say this I spatially mark out; with my leaning; the body knowing and the mind not knowing.

I continue to seed the idea of doing things in a trance by talking about how my Granddad could probably play the piano in his sleep. I also suggest that conscious interference can stop this working just by having that conscious awareness; what is wanted is to just be in the moment and let it happen without thinking about it.

> C: That sucks yeah, but you're right you get similar sort of feelings like when you're trying to draw or copy something and it doesn't do what it's supposed to do, I'd like to er, I don't know, say get a Disney video and be able to just put it on the side and just copy the video, that would be cool, I know people that can do that really really well and I can't understand why they can do it and I can't
>
> D: Yes so it is possible obviously for your body to do things that you don't have an awareness of and obviously consciously you take in a certain amount of information, or at least you

don't take it in, you take it in unconsciously and you're drip feed a certain little amount at a time and a very small amount of information, so you don't notice the fine detail, you don't really sort of, you just let the hand do what it is doing then it doesn't really matter what your conscious mind is thinking, what's on your conscious mind because your hand can just do it…any way that was just an aside…so you haven't been in a deep trance for a while?

C: No

In response to hearing my story the client agrees with what I have said and then states a goal that he would like; that would demonstrate success. He also says that he knows people that can do what he wants to do; which means he knows that it can be done. He also states that consciously he doesn't know how to do this; we are hoping to get him doing this without conscious awareness of how he does it; we just want him to do it.

My response to his statement was to focus on the unconscious process we are laying down and to link it back with him. He had spoke about 'other people' I wanted it to be him that has these abilities so I speak about his body doing things the conscious mind has no awareness of; again I do this by marking out the conscious and unconscious so that when I talk to different locations he will know I am communicating with him on a conscious level or an unconscious level. I then very clearly state what will be expected; that he will notice fine detail and that what his conscious mind is doing has no relevance on his unconscious mind drawing as long as the conscious mind just lets it happen unhindered.

I use the term 'obviously' to convey the meaning that 'this is common knowledge, everyone knows this' this is to depotentiate any resistance by using a term that is unlikely to be questioned. Generally people don't question something that they feel they should know and they often just let it go in because you are not asking for a response. When I work with smokers I often do similar; I will say 'obviously you know all the dangers of smoking, so I'm not going to tell you that…' then I go on to list a load of dangers without receiving resistance because I have framed the statement as just telling them what I am not going to tell them.

Induction

D: There was something I wanted to do and the question is do I do it now or do I not? But I might give it a go…

C: OK (looking confused)

I mention the induction I am going to do in this way as the client has been hypnotised by me before and I didn't want him to know what to expect. I wanted to begin to indirectly hypnotise him. The only way for the client to respond to what I said was for him to go inside his mind to wonder what I am thinking of doing and be confused because of not knowing what to expect. Most effective inductions have an element of confusion to disrupt the conscious mental set as the only way to respond to confusion is to try to find a way out; which will either be by following a clear suggestion by the Hypnotherapist or by going inside their own mind to escape the confusion.

D: Mmm…only because it's just something different for a change…

C: nice day for a change…

D: Yep I was just thinking exactly the same thing….right put your hand on my hand…OK…in a minute I'm going to tell you to push down, now I don't want this to surprise you…so I'm going to tell you in advance what's going to happen…

C: Alright

I mentioned 'something different for a change' so that the client would think of the sentence 'nice day for a change' which implies change happening. I also imply that because I am doing something different that will create the changes we will get in this session.

I tell the client what to do at the start of the induction I will be doing with him and even though I am about to do a very direct induction I still use indirect methods; I tell him I don't want the induction to surprise him and that because of this I will tell him in advance what will happen. I could have just told him directly 'when I do this you will go into a trance like this…etc' but he may have chosen to resist me if I did that so by framing it that I am telling him for his benefit so that he knows what to expect he is more likely to listen to hear what is going to happen rather than feeling that he is being told what to do and feeling manipulated. I then change direction so that he doesn't have too long to analyse what I have just said in case he picks it apart and realises what I am doing. I don't just change to any old subject, I change to the issue of trance, getting some agreement, creating some confusion and again getting the client to recall a previous trance so that when I start the induction he is already partially in the recalled trance and mildly confused making him more receptive and has agreed with me a few times enhancing his receptivity and making him more likely to continue to accept what I say.

D: And you don't mind going into a deep trance do you?

C: Nope

D: You really don't mind?

C: No

To make these sentences more powerful and less likely to be resisted I have used the 'Reverse Yes Set'. The reverse yes set is where you get a no but it is still in agreement with what you have said. By framing the questions negatively the client is less likely to feel the need to disagree. By asking 'You really don't mind?' The client is likely to go inside their mind wondering is there supposed to be a reason to mind? So this causes slight confusion for the client because why would they mind? While they are inside their mind searching previous trance experiences to make sense of the question and see if any previous deep trance experiences should lead them to a different answer they are putting themselves into a deeper trance just before I start the actual induction.

D: What's the deepest trance you've been into? Maybe the one where you lost track of time, although you knew exactly what the time was because you came out of the trance at the right time…Right what I'm going to do is tell you to look into my eyes…not yet…so what I'm going to do is tell you to push down on my hand (client starts pushing) in a minute…in a minute…(client laughing)…and…then when I say sleep (clicking my fingers)…sleep…and do it your own way…you can go relaxed, feet up, head down, your choice, you can slump, do whatever you want, however you want to go into it…OK…**you happy with that.**

C: Yep

I then immediately ask 'what's the deepest trance you've been into?' And then begin to describe a trance experience suggesting that the client can go into a deep trance where he stops paying attention to conscious references like keeping track of time, where he can just be in each moment; and that he can come fully back when he is supposed to. I do this to keep the client occupied with thoughts of being in a trance. He is already partially recalling a trance experience from the last questions I asked; now he will be beginning to recall a specific deep trance experience.

I then jump straight to describing what will happen with the current induction. All of this jumping around keeps some confusion there and keeps the client frustrated needing to become reliant on me for clues about what he is supposed to be doing. The client demonstrates that he is already becoming very receptive by responding literally to my description. He looks into my eyes when I mention it; and pushes down on my hand when I mention that.

> D: Right push down on my hand…really hard…push, harder, keep pushing, harder, harder, keep pushing harder, harder, push down harder, harder, look into my eyes, keep pushing down, harder and harder, harder and harder
>
> C: I can't
>
> D: Keep pushing harder and harder, close your eyes and sleep…

This induction is a very typical rapid induction. It is done very directly and forcefully. While I am doing this induction I am closely watching the client. I am watching his face and shoulders looking for changes that will signal that he has gone inside his mind because it is at that point that I want to give the command to go inside his mind and sleep. After about 30 seconds the client says; just audibly; 'I can't'. It is at this point that I want to offer him a firm suggestion to sleep. When he says he can't he has almost given up trying; he has become slightly confused as how can he push harder if he is pushing as hard as he can. When people are confused they will usually accept any firm suggestion so long as it doesn't go against their personal values. In this induction that is when I say just once more 'keep pushing harder and harder' just to add to the confusion; then I say firmly and directly 'close your eyes and sleep' at the same time as clicking my fingers (which will trigger the reorientation response which is the response that gets fired to alert us to a stimulus making someone focus on just one thing; this fires as we enter dreams which is what can give the sensation of falling); I also pull my hand out from under his which also causes shock creating a trance state. With the shock from the click and from the hand being removed and with the confusion and the direct command the client was very unlikely to not go into a deep trance.

The client slumps back into the chair as if all his muscles have relaxed and his arm falls to his lap.

Trance Deepener

D: That's it, just allow yourself to go down deeper and deeper…that's it, deeper and deeper, just taking deep breaths, that's it……that's it…

I now deepen the trance very directly by telling the client to 'go down deeper and deeper' and saying 'that's it' on the clients out breaths. I also say 'just taking deep breaths' the reason for this was that by saying 'just' implies he is only going to take deep breaths not any other types of breaths.

D: And in a moment I'm going to lift up that hand (looking over at the clients hands), and when I do I'm not going to tell you to put it down, any faster…that's it…than your unconscious mind begins…that's it…to get a sense of what it's like to walk down a flight of stairs…

I continue to frustrate responses to help continue to deepen the trance by being ambiguous; not telling the client which hand I will lift; by saying in a moment not specifying exactly when. I tell the client indirectly that I want him to do an arm levitation by telling him that 'I'm not going to tell you to put it down'. I follow this up with 'any faster' which implies that it will lower by itself (which as the sentence continues links the lowering with the unconscious walking down stairs).

I continue to say 'that's it' on each out breath by the client as I am still deepening the trance so I want to use everything to do this.

> D: And I'm not going to know where those stairs are…I'm not going to know whether they are in a building or whether they are going down to a garden…that's it…or down to a beach…**only you know**…that's it…where those steps are…

I emphasis me not knowing as this implies someone must know and if it isn't me it must be the client. I then tell him that he knows. I find that often it is best to indirectly imply something and lay down a pattern before saying something directly so that you are constantly communicating on two levels; one with implication and metaphor and patterns and the other being direct and able to be understood consciously. Even if I am talking to the unconscious mind I know the conscious mind is listening to some extent so it needs to have a message to follow that seems straight forward enough to not feel a need to analyse too much. Likewise when I am talking to the conscious mind I know the unconscious is listening.

Contingent Suggestions, Compound Suggestions & Nominalisations

> D: And the conscious part of you that is *normally* (emphasised) at the front of your mind can just distract itself in some way as you go down those stairs and it can find its own way of becoming more and more distracted with each step it takes

Contingent suggestions are suggestions where two parts of a communication are linked with an action phrase like; before, during, after, while, as. The two parts don't have to genuinely be linked or relate to each other. In the sentence above I link the conscious mind becoming more distracted with walking down the stairs in the mind even though there is no real link other than me saying 'as'. I also include a compound suggestion. Compound suggestions have two parts of a communication linked with an 'and' or a pause. They imply that because the first part happens so will the second part. In the sentence above I use 'and' to link the going down the stairs with becoming more and more distracted. In effect I am saying the same thing twice in the same sentence using two different language patterns.

In this sentence there is also the implication or presupposition that the client will be walking down the stairs and that he will be consciously distracted. The first half of the sentence was saying the conscious part can distract itself, the second half was about 'how' this distraction will happen. The idea behind this is that the client is now more likely to focus on the 'how' rather than whether the distraction will or will not happen. There is also the implication that the client is in a different state and things that happen in this state are different to normal waking state by emphasising 'normally' implying this isn't 'normally'.

There are also many nominalisations just in this one sentence. Nominalisations are words with no fixed meaning. The listener makes up their own meaning when they hear the words. From the start every session is full of nominalisations by the client and therapist. In the sentence above there is: normally; own way; conscious part of you; front of your mind; distract itself; those stairs (no description of the stairs); more (doesn't specify how much more so the client has to figure that out)

> D: And your unconscious mind that's *normally* at the back (separating conscious/unconscious with change in tonality, change in position my voice is coming from) can come to the front...and you can have an overwhelming sense of...**fully moving to the front**...with each step...that's it...and that unconscious part of you can...**increase in awareness**...of me of what I say...of the way that I say things...of tonality, subtle changes and can really...**fully become aware**...to the front of your mind...

Throughout the induction I am using the word 'and' to make sure that every part is linked. I am separating the conscious and unconscious as I talk to the client so that just by talking in a specific way and having my voice coming from a specific location the client will know whether I am talking to their conscious or unconscious mind or both. I am continuing to embed suggestions and commands to help the client's unconscious mind become more responsive to the way I am communicating to pick up on the marked out communication; including communication that is targeted at the unconscious mind like metaphors and stories.

D: Your conscious mind won't be completely to the back of your mind until you reach the tenth step…that's it…and the unconscious part of you won't completely be at the front of your mind…won't **completely take over awareness** until you reach that tenth step and you can be curious to discover what is at the bottom of those stairs,…that's right…that's right…(I lift up the right arm) and that arm can lower down only at the rate and speed that **you go deeper**…and you won't go all the way down into a deep comfortable trance state until that arm goes all the way down…that's it…all the way down…that's it…that's it…(the arm has lowered now)

The main purpose of the steps is as a deepener and to create greater separation between what is coming up which is the laying down of the patterns for improved artistic ability and the initial conscious thinking. Normally when people use steps they count the client down the steps. This relies on the client going at the speed the therapist sets. By doing the arm levitation I have a signal that I can visually observe that is linked to the client going deeper at a speed they choose. I continue to use negative phrasing saying 'you won't … until…' this generally creates less resistance and is slightly harder for the client; especially when they are relaxed; to unpick and analyse. Using negatives in this way people often hear the first part of the sentence (for example: You won't go all the way down into a deep and comfortable trance state) and if they are going to resist they often respond by doing it now and responding opposite to the statement. It is like a reverse double bind in that they can resist and respond now; or they can follow the suggestion and respond when they are asked to. The issue is about time not about whether they will do what they are being asked or not. It takes a lot for a client to then unpick the sentence and decide they will actually ignore it completely and not do anything. They are far more likely to choose - go against the therapist and do it now; or go with the therapist and do it when I am supposed to. Most people in therapy have made a commitment to be there to receive help so doing nothing when asked to is often not on their mind when they have an alternative to feel a sense of control and whatever other needs they would meet by going against the therapist.

D: And you can wonder where you are going to wonder next…and somewhere there you can discover a painting…and

I don't know what that painting will be of and what it is that **it can teach you something about yourself**...

Now that the client has reached the bottom of the steps I add a little confusion by using words that sound similar with one talking about thinking and one about action. I use many nominalisations as I don't know where they are going to be in their mind. And because I don't know where the painting that I want in their mind is going to be I don't tell them where it is; I give them the option to discover it. I keep using the word 'and' to continually compound end statement on the previous statement so that they build on each other. In many sessions I have observed people will say things like '...and in front of you, you will see a picture...' If the person can see in front of themselves in their minds eye before you have said that and there was no picture then you will mismatch their internal reality. By suggesting 'wonder and somewhere' it leaves the painting to be found. I also state a truism 'I don't know what that painting will be of...' This is undeniable with the implication that they must know. Being a truism they again will be in agreement with me. It is also followed with a poorly formed sentence. I run one sentence into the next using the words 'it can teach you' to transition from the end of a sentence with one meaning - saying ...what it is that it can teach you - then continuing the sentence giving a suggestion of what I want that painting to teach - …it can teach you something about yourself. There is so much going on in the sentence with thinking about the wondering, finding a painting, hearing truisms that create agreement; that to also keep track of the final embedded command is difficult it just sinks in because analysing that as well takes considerable effort.

> D: And there can be something curious about the painting…and as you pay all your attention to me so you can notice the things that I say and the things that I do and perhaps it will be the way that I say things…that's it…and I wonder what it is that's curious about the painting…

I am continuing with many nominalisations like 'curious' 'attention' 'things that I say' 'things that I do' 'way that I say things' 'wonder'. These help to keep the client on an inner search for their own meaning to what I am saying and doing. I sandwich the paying attention to me and emphasising to the client that there is something important in what I am saying and doing between two statements about being curious about the painting. This is partly to create amnesia for this paying attention. I want the client to unconsciously be responsive but to feel these bits haven't been said; that they were just listening to me talking about the painting.

Open Ended Suggestions

> D: Could it be that movement and I wonder where that movement is and whether it is a little bit of movement or a lot of movement and whether that movement is in the centre or off to the sides or round the edges…and I wonder what type of movement it is whether it's a sort of wavy movement or a swirly movement or some other kind of movement and whether it has a 3d effect or a 2d effect…

While the client is paying attention to the painting I want to generate movement. I don't want to mismatch the client's inner reality (which could include movement or no movement in the painting). I decided to be fairly strong on saying that there is movement so I say there is movement but because I know the client could be seeing a still image I want to cover all bases with suggestions that lead to the movement being their but perhaps not at first observed. I mention the location and give the option for it being a little bit or a lot and the type of movement. I don't want to give a chance for the client to stop and think there is no movement so I say all of these options in fairly quick succession. I'm always implying there is definitely movement there even if at first it was so small it wasn't noticed until I mentioned it.

> D: That's it and you can be curious to **pay attention**...to whether there are sounds there and perhaps they are coming from the picture or from elsewhere and maybe behind you or in front of you or to the left or the right or from above or maybe below…that's it…that's it…that's it…

This is a technique I use quite often; where I will suggest an idea as if any possibility is an option then follow a preferred route. For example above I suggest paying attention to see if there are sounds there; giving the choice that there may be silence but before the client has time to think about it I suggest that sounds are there it is now a question of where they are coming from. I then go on to give multiple options of where the sounds could be originating.

> D: And another curiosity about this picture is that **you can step inside this picture**…that's it…and I don't know what it's like the other side of the picture…

When I said 'step inside the picture' I could see the client had done that by changes in his physiology which I acknowledged by saying 'that's it' I then followed this up by immediately stating a truism that I don't know what it is like the other side of the picture again implying that they do. And for the client to know what it is like they have to be there. By having the client step into the painting it takes them even deeper into trance. Any time you have someone change where they are in their mind they go deeper and layer their trance experience so that the last place they go gets sandwiched between the previous places, which are sandwiched between the places before that etc… This leads to usually spontaneously getting amnesia for the deeper parts of the trance.

Metaphors for Unconscious/Conscious Processes

D: That's it…you know *normally*…that conscious part of you…guides your decisions…guides what you do, what you're thinking about…it's a bit like a driver of a train…the driver is just a small part of the whole thing…and the driver can just see what is outside his window…and the driver knows that there are 8 carriages behind and the driver knows that each of those carriages is full of people and that each of those people are saying their own things and doing their own things and each of those people are in control of what they're in control of…some are reading newspapers…some are listening to music and some are planning ideas and many of them are lost in thought…that's it…yet all the driver is aware of even though the driver knows all of that is there…is what's through the window…

I have used a metaphor to parallel the conscious mind by having the conscious mind as the driver of a train and the passengers as neurons that are all doing their own thing independent of the driver. People consciously know their mind is doing lots of things at once that they have no awareness of; and that all they are aware of is what they consciously are currently aware of. Like the driver only being able to see out the window. I am conveying this message as a metaphor as this is an easier way to lay down a pattern for the unconscious mind to use.

D: That's it…whereas the unconscious part of you is like a super being floating above the train like superman flying above the train where he can see the passengers…because he can see through the walls because he can see what they are all doing, he can notice their behaviours their language...he can fly down and talk to them…he can even make them change their behaviours…he can ask someone to stop reading their newspaper and they would stop…he could ask someone to stop listening to music and they would stop…he could interrupt someone having a conversation and they'd forget what it was that they were talking about…

To describe the unconscious mind I use the term super being as it has many positive connotations; I describe some of the strength of the unconscious mind and what it can influence by talking about how it can influence the passengers. The idea I want to convey is that a conscious knowing and an unconscious doing are two different things. I want the client to be able to 'just do' on an unconscious level.

Post Hypnotic Suggestion

D: That's it…now you can be curious…as to how you are going to use all of your unconscious resources…how you are going to use all of the talents you've got that are normally held back behind the doors of the conscious part…you can be curious about what the improvements will be…and how your mind and your circuitry of your brain will make aspects of the improvements permanent in a comfortable way…that's it…that's it……and you can get a sense of what it is like to see you in your mind…to see you in the future…

There is implication running through this section. When you use implication or presuppositions they act like post hypnotic suggestions. By saying 'you are going to' and 'will' it is placing what I am saying in the future as a certainty rather than a possibility. I follow the suggestions up by building on each preceding suggestion; so I firstly state what will happen in the future, then I move on to 'see you in your mind' then onto a context for that 'you' that is being seen - in the future. If I didn't add that the client could see themselves at any age and any time even made up ages and times (like in the distant past or a futuristic world). I want to keep what I say as unthreatening as possible so I build one thing on another hoping to go just slightly faster than the client's awareness of what I am suggesting so that I am leading their internal reality now. In the same way that you can wear glasses and not notice until someone mentions them; I want the client to assume they've just not noticed something until I mentioned it; rather than it not previously being a part of their reality.

D: And you remember Superted…Superted used to say his magic word and he would change from an ordinary teddy bear into a super-teddy bear…and consciously you've watched Superted and you could never quite hear the word…you'd never know consciously what the word is no matter how often you watched it…and you can see yourself…and you can see that you having a code word…and you can watch yourself say that code word to yourself…you can watch yourself say that code word to yourself….

I now use the cartoon character Superted to introduce the idea of a code word for triggering the artistic ability. Ideally I want the code word to be something internal and unconscious rather than something the client has to consciously say. It is like when a hypnotist sets up the word 'sleep' to re-induce trance. I want the client to have a word to re-induce artistic abilities but I want it to come from the client's unconscious mind. To cement ideas and reprogramming I normally give the idea; then have them watch themselves doing the new behaviour; then have them go into that version of themselves to experience actually doing what they have just watched themselves do. This is generally a very effective way of getting that behaviour into the client's future. It also matches the way people normally do things. They get the information (me giving the idea); then think about what they are going to do (see themselves doing the behaviour); then they do it (stepping into that future them and experiencing it). If you just jump someone into just doing a new behaviour it may not stick because there was no planning or mirroring reality. Also by doing it this way if there are any problems with the way they see things go it can be changed to be just right before they go into the experience.

> D: And when you watch yourself becoming overwhelmed by that compulsive artistic ability…and I don't know if it's thousands of times or even tens of thousands of times or even realistic and lifelike…in comparison to how the conscious part of you…that other part of you does art…that's it…that's it…and you can watch and I wonder what you see…and I wonder what you see how well you're doing that art…does it look haphazard to start with like when you watch Rolfe Harris…where that other part…the unconscious part has its own way of drawing where it takes control…**it takes control**…and I wonder whether that you reports that it's like the hand doing all of that work themselves…whether it's like

> an image just being printed onto a page straight from the mind…whether it's like the hands just get a compulsive feeling…whether it's like the hands get a compulsive feeling to carry out that artistic talent…

People with 'natural' artistic ability or people that are 'naturally' talented at anything generally have a high level of compulsion to do what they are interested in. In this section it is this compulsion that I am working at installing. I offer lots of ideas about how that compulsiveness will take effect and that it comes from the unconscious part of the client not the conscious part. I do this by taking about the hands doing the work and that the unconscious mind has its own drawing style. The presupposition through all of this is still one of that improved drawing ability taking place; it's now a question of what it will be like to the conscious mind as an observer rather than whether it will happen. I am also talking to the client as someone that is still observing that future them. They are not yet in that future them.

> D: And I wonder how long that lasts…is it for 20 minutes or half an hour or is it for a full hour…and I wonder whether it ends because you decide it's time to stop or whether it ends

> because the time is up…and I wonder how long it lasts…is it twenty minutes, half an hour or maybe even a full hour…or somewhere in-between…

I now limit the duration of that compulsive behaviour. I don't want it to carry on indefinitely because that could have negative effects on work and family life. I want the client's unconscious mind to decide the duration so I offer choices around how long it will last rather than just telling the client how long it will last. Whenever doing any work with clients the therapist needs to be mindful of the positive and negative effects of making the change. By having the client view the changes first it gives them the opportunity to see if the changes are acceptable. If they are not you have given the client the option to decide not to go into themselves when we get to that stage and also to make any changes now before they do. It is a bit like starting with hindsight; the client can see if the changes are acceptable from a dissociated position. Being dissociated gives a useful view on things. Like watching a football match and knowing a player should have passed because you saw another player open; yet the player that should have passed wasn't in a position to see what you could see so made different decisions.

D: And like all resources…**you can use**…the skills and talents and you can transfer to other areas…and that's TRANCEfer to other areas…that's it…excellent…and you know what it's like to…**step inside that you there and experience that deep compulsive desire to create photorealistic art work**…and I wonder what changes happen at a neurological level that are completely comfortable and healthy that build the talent and increase the talent by temporarily numbing down an area of the brain and I wonder how that numbing down takes effect…

I use the term TRANCEfer to imply going into trance to transfer skills and abilities.

Shutting Down Perceptual Filters

D: You can be curious as to whether it will be like certain signals not being allowed through or certain signals taking a different route around the brain…performing those talents with those signals taking different route to how you perform it consciously and you can imagine what it is like to be that you there carrying out…that's it…that artistic compulsion… and I wonder what it feels like…do the hands feel tingly…or is it in the arms or is it on the back of the neck…

I want to generate feelings associated with the specific trance of improved artistic ability so I ask what it feels like then I tell the client in the next sentence what it will feel like; that it will be a tingling. I them make the focus on where that tingling will be rather than if there will be a tingling or not. I am also setting up the next part of the session where I want the tingling to be link to energy.

D: You can be curious as to how that compulsion takes effect once you hear that word and I don't know whether the word is Leonardo, or whether the word is Rafael or some other famous artist…that's it…that's it……that's it…and you can go deeper…and deeper…and you know what somnambulism is……that's right…and when that artistic desire…compulsion takes effect I wonder whether it makes the fingers twitch with nervous energy desperate to carry it out or whether that doesn't show…

The self-hypnotic suggestion is bought back in again here and linked to the feeling/energy. I now make the focus on whether the energy will be visible to others or not rather than if it will be there or not. When I introduce something I want to be there I like to make the focus on something to do with that new thing rather than if that new thing will be there or not.

> D: And you can imagine in your mind a panel and it's a panel with many levers on…and each lever is logically labelled…that's it (he moved)..and you can imagine turning up that lever to artistic ability…putting it up to full…that's it…that's it…and in a moment I'm going to lift up your right hand and when I do I'm not going to tell you to put it down any faster than you…**become absorbed**…in the idea of being a great artist…and I don't know whether it's going to be an artist that is going to be a mixture of many artists or an artist that is going to be greater than any artist you've ever known…that's it…and I don't know whether your hand coming down will take two minutes, three minutes or five minutes or somewhere in between…and I don't know how long that will be on the inside…for you…and I wonder what you get up to while that is happening……

The client already has the experience of using arm movement to go deeper into a trance which is becoming more absorbed. I now use it again to become absorbed in the idea of becoming a great artist. Again I introduce an idea; the idea of becoming deeper absorbed; then I focus attention on time rather than whether the absorption will happen or not and then change focus to what the conscious mind will be getting up to.

> D: And when it's time to come out of a trance and come all the way back to the room you can come back with the artistic abilities…and I'll let you know when that time is…and I wonder what it will have felt like to have those changes occurred on a deep neurological level…first changes can occur though all the connections, pathways, circuitry…that's it…and on one level when you come back how you can be curious as to where those changes came from and how those changes occurred also…

The suggestions given here are given very directly that when the client comes out of the trance they will come back with the artistic abilities. I use the word occurred rather than occur in the middle of a sentence to imply they have happened; then I mention a curiosity about where those changes have come from to again change the focus from whether it happens or not to trying work out where the changes came from.

D: And I wonder whether you will be amazed or at least shocked...and you know the studies that have been done by shutting down areas temporarily in a healthy way...whether it is shutting down or just not allowing the signals to get through temporarily from the logical rational hemisphere...part of the brain...the creative part of the brain the part that notices fine detail notices every little thing, notices millions and millions of bits of information every single second...that part of the brain can take control...and then it will end up with you having...(client moved his arm to scratch)...that's it...many abilities...that's it...that's it...now as you're going to achieve great things...that's it...that's it...we want you to take your time to do this...**take your time to do this now**...that's it...that's it...(I lift the clients arm above his head)...that's it...that's it...and on many levels you can begin to count backwards from 100 and let the unconscious work...that's it...and you know what your right foot feels like and I wonder if it feels different from the left...that's it...that's it...and your left hand can be left there lightly resting where it is and I wonder how your eyelids feel…that's it…that's it…making this completely…

While the client has their arm in the air and is counting down in his mind I want to jump his attention all over the place to stop him focusing on the lowering arm and the internal unconscious work.

D: That's it and you know a minute of my time can seem like a longer time of yours just like you can experience…that's it…(arm fully came down)…a longer time that just goes by from minute to minute…and you can take a minute of my time to go deep and comfortable inside your mind into a deep and comfortable focused state of mind…allowing that artistic ability to develop and enhance itself and the more it enhances itself the more pleasure you can experience inside your mind and you can take a minute of total silence to do that now…(minutes silence - I just sit observing minimal cues)…

I sit and observe and watch for signs of increasing pleasure to see that he is enhancing his ability.

> D: That's it…that's it…and you know you can control blood flow, you can make yourself blush on half of your face and if you get a cut you can make it so that the blood stops flowing over the cut any more than is necessary to keep the wound clean…you can make an arm numb or a leg numb or half a head numb…you can increase your metabolism or slow it down…you can alter any system in your body with your mind…

To link back with the earlier statement of making part of the brain numb I sandwich the suggestion for this again in the middle of a collection of truisms about what the mind is capable of.

> D: Now when it's time for you to open your eyes I'll ask you again to draw a horse and again you will get one minute to do it in…and you can be curious as to how much better that horse will be drawn…will it be lifelike, will it be hundreds or thousands of times improved on the last one…how much can you manage to draw in one minute…you can be curious as to how much you can draw of that horse in one minute and the

level of detail you can draw…and you can always keep in mind that ability those abilities you have…and you can always keep in mind how to get that artistic ability…and you can always keep that in mind……that's it…and you can open your eyes now…

My use of the term 'you can always keep in mind' is specifically used because if you always keep something in mind it means it is always there. It is a post hypnotic suggestion to make everything stick.

D: (Now talking very normally) Right want to have a bash at drawing the horse again…we'll see how it goes now…help if the pen works (pen not working so the client changed pen)…(one minute given to re-draw the horse)…times up…how do you think you did?

C: It looks more like a horse than the last one did; it's more proportionate I think

D: Yes, drawing style was different as well

(client starts to carry on drawing)

C: I'm not supposed to be doing this now am I?

D: No, I'll show this to the camera before we...if you do start finishing it off or something

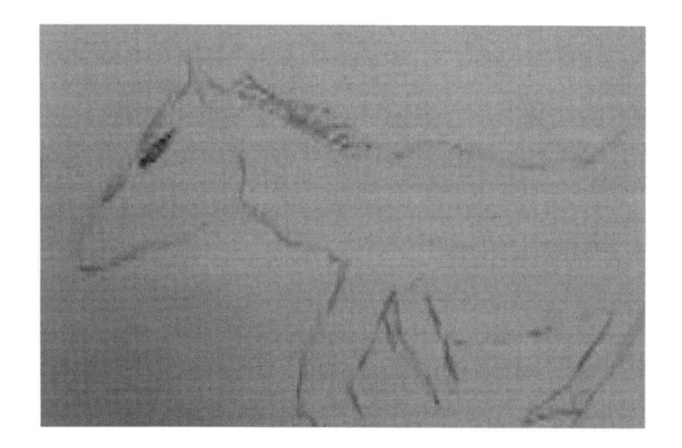

C: That is quite different isn't it (client picks up a pen again to carry on drawing)

D: And that's only a first thing drawn after being zapped

Next few minutes was spent me just watching as the client compulsively kept drawing, putting the pen down to talk then carrying on getting absorbed in drawing again, getting more involved in what he was doing and adding more and more detail, then adding colour and motion. In total he spent about 4 minutes on the picture adding to it, putting the pen down for a few moments to talk but then not talking instead he would pick the pen up and carry on drawing. I tried to get his attention to discuss his experience but he struggled to tell me as he was more absorbed in continuing to draw. Below is the picture after about four minutes of drawing on it.

After the session the client went home and drew an image off of a Disney video case. He said it only took him a few minutes to sketch it. He then put it on his computer and coloured it in. He said he was amazed at what he had done and didn't understand how it was possible.

Index

Printed in Great Britain
by Amazon.co.uk, Ltd.,
Marston Gate.